Student Solutions M

Fundamentals of Algebraic Modeling

SIXTH EDITION

Daniel L. Timmons
Alamance Community College

Catherine W. Johnson
Alamance Community College

Sonya M. McCook
Alamance Community College

Prepared by

Daniel L. Timmons
Alamance Community College

Catherine W. Johnson
Alamance Community College

Sonya M. McCook
Alamance Community College

BROOKS/COLE
CENGAGE Learning·

Australia • Brazil • Japan • Korea • Mexico • Singapore • Spain • United Kingdom • United States

BROOKS/COLE
CENGAGE Learning·

For product information and technology assistance, contact us at
Cengage Learning Customer & Sales Support,
1-800-354-9706

For permission to use material from this text or product, submit all requests online at **www.cengage.com/permissions**
Further permissions questions can be emailed to
permissionrequest@cengage.com

ISBN-13: 978-1-285-42042-4
ISBN-10: 1-285-42042-X

Brooks/Cole
20 Davis Drive
Belmont, CA 94002-3098
USA

Cengage Learning is a leading provider of customized learning solutions with office locations around the globe, including Singapore, the United Kingdom, Australia, Mexico, Brazil, and Japan. Locate your local office at: **www.cengage.com/global**

Cengage Learning products are represented in Canada by Nelson Education, Ltd.

To learn more about Brooks/Cole, visit **www.cengage.com/brookscole**

Purchase any of our products at your local college store or at our preferred online store **www.cengagebrain.com**

Printed in the United States of America
1 2 3 4 5 6 7 16 15 14 13 12

Table of Contents

Chapter R   1
Practice Set R-1   1
Practice Set R-2   2
Practice Set R-3   6
Practice Set R-4   8
Review Problems   9
Chapter R Test   12

Chapter 1   15
Practice Set 1-2   15
Practice Set 1-3   18
Practice Set 1-4   21
Review Problems   25
Chapter 1 Test   28

Chapter 2   33
Practice set 2-1   33
Practice Set 2-2   36
Practice Set 2-3   37
Practice Set 2-4   41
Practice Set 2-5   42
Review Problems   43
Chapter 2 Test   45

Chapter 3   49
Practice Set 3-1   49
Practice Set 3-2   51
Practice Set 3-3   55
Practice Set 3-4   59
Practice Set 3-5   63
Review Problems   66
Chapter 3 Test   70

Chapter 4   75
Practice Set 4-1   75
Practice Set 4-2   77
Practice Set 4-3   79
Practice Set 4-4   81
Practice Set 4-5   85
Practice Set 4-6   89
Review Problems   91
Chapter 4 Test   98

Chapter 5   103
Practice Set 5-1   103
Practice Set 5-2   104
Practice Set 5-3   105
Practice Set 5-4   109
Practice Set 5-5   110
Practice Set 5-6   112
Practice Set 5-7   112
Practice Set 5-8   113
Review Problems   115
Chapter 5 Test   117

Chapter 6   121
Practice Set 6-1   121
Practice Set 6-2   121
Practice Set 6-3   129
Practice Set 6-4   136
Review Problems   139
Chapter 6 Test   146

Chapter 7   151
Practice Set 7-1   151
Practice Set 7-2   152
Practice Set 7-3   153
Practice Set 7-4   155
Practice Set 7-5   156
Practice Set 7-6   158
Practice Set 7-7   159
Practice Set 7-8   160
Review Problems   163
Chapter 7 Test   165

Chapter 8   167
Practice Set 8-1   167
Practice Set 8-2   168
Practice Set 8-3   170
Practice Set 8-4   172
Practice Set 8-5   174
Practice Set 8-6   175
Practice Set 8-7   178
Review Problems   181
Chapter 8 Test   183

TI-84® Quick Reference Guide
Graphing Points with the Stat-Editor   185
Solving Equations By Graphing   187
Graphing Piece-Wise Functions   188

Solutions to Practice Problems in the Text

Chapter R: A Review of Algebra Fundamentals

Practice Set R-1

1. $\{-2.6, -\sqrt{5}, -1, 0, \frac{7}{13}, 3\frac{1}{2}, 4\}$

3. a) $\sqrt{7} < 3$ b) $-5 > -5\frac{1}{4}$ c) $\pi > 0.3$

5. $5 + (-3) = 2$

7. $-6 + (-5) = -11$

9. $60 \div 0 =$ undefined (division by 0 is impossible)

11. $|6 - 8| = |-2| = 2$

13. $-1\frac{1}{3} + \frac{7}{8} = -\frac{4}{3} + \frac{7}{8} = -\frac{32}{24} + \frac{21}{24} = -\frac{11}{24}$

15. $(-2.5)(-1.6) = 4$

17. $\left(\frac{1}{4}\right)\left(-\frac{3}{8}\right) = -\frac{3}{32}$

19. $-(-2) + 10 = 2 + 10 = 12$

21. $-\frac{8}{9} \div \frac{4}{21} = -\frac{8}{9} \times \frac{21}{4} = -\frac{168}{36} = -\frac{168 \div 12}{36 \div 12} = -\frac{14}{3}$

23. $7 \cdot 3 - 2 \cdot 13 = 21 - 26 = -5$

25. $\dfrac{15}{[-16 - (-11)]} = \dfrac{15}{-5} = -3$

27. $8 - (-6)(4 - 7) = 8 - (-6)(-3) = 8 - 18 = -10$

29. $\dfrac{2(-6 + 6)}{23 - 97} = \dfrac{2(0)}{-74} = \dfrac{0}{-74} = 0$

31. $(-8)^2 - 7(8) + 5 = 64 - 7(8) + 5 = 64 - 56 + 5 = 13$

33. $\dfrac{2^2 + 4^2}{5^2 - 3^2} = \dfrac{4 + 16}{25 - 9} = \dfrac{20}{16} = \dfrac{5}{4}$

35. $\dfrac{3(-5 + 1)}{12(3) + (-5 + 2)(-3 - 1)} = \dfrac{3(-4)}{12(3) + (-3)(-4)} = \dfrac{-12}{36 + (12)} = \dfrac{-12}{48} = -\dfrac{1}{4}$

37. $\dfrac{5}{8} - 5\left(\dfrac{1}{8}\right) = \dfrac{5}{8} - \dfrac{5}{1}\left(\dfrac{1}{8}\right) = \dfrac{5}{8} - \dfrac{5}{8} = 0$

39. $1.25 + \dfrac{6.5}{0.5} + (0.25)^2 = 1.25 + \dfrac{6.5}{0.5} + 0.0625 = 1.25 + 13 + 0.0625 = 14.3125$

1

41. $\dfrac{3^3 - 2^3}{-4(-3+1)} = \dfrac{27-8}{-4(-2)} = \dfrac{19}{8} = 2.375$

43. $-4[(-2)(6) - 7] = -4[(-12) - 7] = -4[-19] = 76$

45. $(3 - 8)(-2) - 10 = (-5)(-2) - 10 = 10 - 10 = 0$

47. $\dfrac{-\left|-8\right|}{-2} = \dfrac{-8}{-2} = 4$

49. $|-4| = 4$, the opposite of 4 is -4, $4 + (-4) = 0$

Practice Set R-2

1. $x + 6 = -4$
    $x + 6 - 6 = -4 - 6$
        $x = -10$

3. $3x + 7 = x$
    $3x + 7 - 7 = x - 7$
      $3x - x = x - 7 - x$
        $2x = -7$
          $x = -\dfrac{7}{2} = -3.5$

5. $2(x - 8) = 4$
    $2x - 16 = 4$
    $2x - 16 + 16 = 4 + 16$
        $2x = 20$
          $x = 10$

7. $35 - 3x = 5$
    $35 - 3x - 35 = 5 - 35$
        $-3x = -30$
          $x = 10$

9. $-4x + 5 = -54 + 41$
    $-4x + 5 = -13$
    $-4x + 5 - 5 = -13 - 5$
        $-4x = -18$
          $x = \dfrac{-18}{-4} = 4.5$

11. $7 - 3x = 2 - 5x$
    $7 - 3x - 7 = 2 - 5x - 7$
        $-3x = -5 - 5x$
    $-3x + 5x = -5 - 5x + 5x$

2

$$2x = -5$$
$$x = \frac{-5}{2} = -2.5$$

13. $3x + 9 = 42$
$$3x + 9 - 9 = 42 - 9$$
$$3x = 33$$
$$x = 11$$

15. $-7(2x + 3) = -7$
$$-14x - 21 = -7$$
$$-14x - 21 + 21 = -7 + 21$$
$$-14x = 14$$
$$x = -1$$

17. $3(x - 2) + 2 = 11$
$$3x - 6 + 2 = 11$$
$$3x - 4 = 11$$
$$3x - 4 + 4 = 11 + 4$$
$$3x = 15$$
$$x = 5$$

19. $3x + 5 = -5x - 8 + 1$
$$3x + 5 = -5x - 7$$
$$3x + 5 + 5x = -5x - 7 + 5x$$
$$8x + 5 = -7$$
$$8x + 5 - 5 = -7 - 5$$
$$8x = -12$$
$$x = -\frac{12}{8} = -1.5$$

21. $-\dfrac{3x}{8} = -\dfrac{15}{32}$
$$32\left(-\frac{3x}{8}\right) = 32\left(-\frac{15}{32}\right)$$ [multiply by the common denominator 32]
$$-12x = -15$$
$$x = \frac{-15}{-12} = \frac{5}{4} = 1.25$$

23. $\dfrac{x}{5} - 12 = 7$
$$5\left(\frac{x}{5}\right) - 5(12) = 5(7)$$ [multiply by the common denominator 5]
$$x - 60 = 35$$
$$x - 60 + 60 = 35 + 60$$
$$x = 95$$

3

25. $w + \frac{3}{4} = \frac{5}{8}$

$w + \frac{3}{4} - \frac{3}{4} = \frac{5}{8} - \frac{3}{4}$

$w = \frac{5}{8} - \frac{6}{8} = -\frac{1}{8}$ [get a common denominator for $\frac{5}{8}$ and $\frac{3}{4}$ and combine]

27. $\frac{2}{3}x + 8 = \frac{1}{2}$

$6\left(\frac{2}{3}\right)x + 6(8) = 6\left(\frac{1}{2}\right)$ [multiply by the common denominator 6]

$4x + 48 = 3$

$4x + 48 - 48 = 3 - 48$

$4x = -45$

$x = \dfrac{-45}{4} = -11.25$

29. $\frac{1}{3} + \frac{1}{6}x - 2 = \frac{2}{3}x + \frac{1}{2}$

$6\left(\frac{1}{3}\right) + 6\left(\frac{1}{6}x\right) - 6(2) = 6\left(\frac{2}{3}x\right) + 6\left(\frac{1}{2}\right)$ [multiply by the common denominator 6]

$2 + x - 12 = 4x + 3$

$x - 10 = 4x + 3$

$x - 4x - 10 = 4x - 4x + 3$

$-3x - 10 = 3$

$-3x - 10 + 10 = 3 + 10$

$-3x = 13$

$x = -\dfrac{13}{3}$

31. $\frac{3}{4}(4 - x) = 3 - x$

$3 - \frac{3}{4}x = 3 - x$ [distributive property]

$4(3) - 4\left(\frac{3}{4}x\right) = 4(3) - 4(x)$ [multiply through by the common denominator 4]

$12 - 3x = 12 - 4x$

$12 - 3x + 4x = 12 - 4x + 4x$

$x + 12 - 12 = 12 - 12$

$x = 0$

33. $k - 76.98 = 43.56$

$k - 76.98 + 76.98 = 43.56 + 76.98$

$k = 120.54$

4

35. $x - 0.5x = 12$
$0.5x = 12$
$x = 24$

37. $3.5x + 3 = x - 1.75$
$3.5x - x + 3 = x - x - 1.75$
$2.5x + 3 = -1.75$
$2.5x + 3 - 3 = -1.75 - 3$
$2.5x = -4.75$
$x = -1.9$

39. $-4y + 3 = 12 - 4y$ Since all of the variables have been eliminated
$-4y + 4y + 3 = 12 - 4y + 4y$ and the resulting statement ($3 = 12$)
$3 = 12$ is false, there is no solution to the equation.
$\varnothing$

41. $7a - (a - 5) = -10$
$7a - a + 5 = -10$
$6a + 5 = -10$
$6a + 5 - 5 = -10 - 5$
$6a = -15$
$a = \dfrac{-15}{6} = -2.5$

43. $3(x - 5) - 5x = 2x + 9$
$3x - 15 - 5x = 2x + 9$
$-2x - 15 = 2x + 9$
$-2x - 15 - 2x = 2x + 9 - 2x$
$-4x - 15 = 9$
$-4x - 15 + 15 = 9 + 15$
$-4x = 24$
$x = -6$

45. $6x - 2(3 - 4x) = 6(3x + 2)$
$6x - 6 + 8x = 18x + 12$
$14x - 6 = 18x + 12$
$14x - 18x - 6 = 18x - 18x + 12$
$-4x - 6 = 12$
$-4x - 6 + 6 = 12 + 6$
$-4x = 18$
$x = -4.5$

47. $4(3 - x) + 2x = -12$
$12 - 4x + 2x = -12$
$12 - 2x = -12$

5

$$12 - 2x - 12 = -12 - 12$$
$$-2x = -24$$
$$x = 12$$

49. $3(x - 2) + 7 = 5 - 2(x + 3)$
$$3x - 6 + 7 = 5 - 2x - 6$$
$$3x + 1 = -2x - 1$$
$$3x + 1 - 1 = -2x - 1 - 1$$
$$3x = -2x - 2$$
$$3x + 2x = -2x + 2x - 2$$
$$5x = -2$$
$$x = -0.4$$

Practice Set R-3

1. $8\% = \dfrac{8}{100} = \dfrac{2}{25}$

3. $180\% = \dfrac{180}{100} = \dfrac{9}{5}$

5. $33\dfrac{1}{3}\% = \dfrac{100}{3}\% = \dfrac{\frac{100}{3}}{100} = \dfrac{100}{3} \div 100 = \dfrac{100}{3} \cdot \dfrac{1}{100} = \dfrac{1}{3}$

7. $5\% = 5 \div 100 = 0.05$

9. $0.05\% = 0.05 \div 100 = 0.0005$

11. $1.5\% = 1.5 \div 100 = 0.015$

13. $125\% = 125 \div 100 = 1.25$

15. $0.005 \cdot 100 = 0.5$ so $0.005 = 0.5\%\%$

17. $\dfrac{2}{3} \cdot 100 = \dfrac{200}{3} = 66\dfrac{2}{3}$ so $\dfrac{2}{3} = 66\dfrac{2}{3}\%$

19. $1.5 \cdot 100 = 150$ so $1.5 = 150\%$

21. 15% of $75 = 0.15 \times 75 = 11.25$

23. 1.5% of $32 = 0.015 \times 32 = 0.48$

25. 2.5% of $x = 33.5$
$$0.025x = 33.5$$
$$x = 1340$$

6

27. 15% of $x = 7.5$

$\qquad 0.15x = 7.5$

$\qquad\qquad x = 50$

28. 1.25% of $x = 0.15$

$\qquad 0.0125x = 0.15$

$\qquad\qquad x = 12$

29. 306 is what percent of 450?

$\qquad 306 = x(450)$

$$\frac{306}{450} = x$$

$\qquad 0.68 = x$

$\qquad 68\% = x$

Using proportions, $\dfrac{x}{100} = \dfrac{306}{450}$

$\qquad\qquad 450x = (306)(100)$

$\qquad\qquad\quad x = 68$ or 68%

31. 0.375 is what percent of 25?

$\qquad 0.375 = x(25)$

$$\frac{0.375}{25} = x$$

$\qquad 0.015 = x$

$\qquad 1.5\% = x$

Using proportions, $\dfrac{x}{100} = \dfrac{0.375}{25}$

$\qquad\qquad 25x = (0.375)(100)$

$\qquad\qquad 25x = 37.5$

$\qquad\qquad\quad x = 1.5$ or 1.5%

33. $\dfrac{x}{100} = \dfrac{18 \textit{ free throws}}{24 \textit{ attempts}}$

$\qquad 24x = (100)(18)$

$\qquad 24x = 1800$

$\qquad\quad x = 75$ or 75%

35. 65% of free throws $= 0.65 \times 12 = 7.8$ or 8 free throws.

37. 18% of the bill $= 0.18 \times \$325 = \$58.50$

39. 26% of weight $= 0.26 \times 180 = 46.8$ lb. of skin

41. 9% of purchase price $= 0.09 \times \$42.50 = \$3.83$

43. $\dfrac{x}{100} = \dfrac{6}{200}$

     200x = (6)(100)

     200x = 600

        x = 3 or 3% sleep more than 8 hours

       3% of people = $0.03 \times 700 = 21$ people out of 700 sleep more than 8 hours

45. 8.3% of Georgia residents aged 25 and older = $0.083 \times 5,185,965 = 430,435$ have professional or graduate degrees.

47. 4% increase $\times$ current pay = $0.04 \times \$7.75 = \$0.31$ increase; new rate = $\$7.75 + \$0.31 = \$8.06$

49. $\dfrac{151.80 - 172.50}{172.50} \cdot 100 = \dfrac{-20.7}{172.50} \cdot 100 = -12$ or a discount of 12%

51. $\dfrac{\$80,000 - \$75,000}{\$75,000} \cdot 100 = \dfrac{\$5,000}{\$75,000} \cdot 100 = 6.\overline{6}$ or approximately $6\dfrac{2}{3}$% increase

Practice Set R-4

1. 230,000,000   (move decimal 8 places right)
3. 3000   (move decimal 3 places right)
5. 0.605   (move decimal 1 place left)
7. 0.003   (move decimal 3 places left)
9. 7.5   (move decimal 0 places)
11. 6.5×10^2
13. 3.2×10^9
15. 5×10^{-2}
17. 7.5×10^{-7}
19. 6.5×10^0
21. 1.47×10^8 km
23. 4.8×10^{-4}
25. 1.2495×10^{-9}
27. 4.375×10^{-10}
29. 4.6865×10^{21}
31. Au's mass = 196.9665, now as per example 4

$$\dfrac{1}{196.9665}(6.022 \times 10^{23}\ atoms) = 3.0573727 \times 10^{21}\ atoms$$

8

Chapter R: Review Problems

1. $-\dfrac{2}{3}, \dfrac{3}{2}$

2. $9, \quad -\dfrac{1}{9}$

3. $-|-3 + 5| = -|\,2\,| = -2$

4. $-7^2 + 9 = -49 + 9 = -40$

5. $6 - 3(4 - 7) \div 4 = 6 - 3(-3) \div 4 = 6 - (-9) \div 4 = 6 - (-2.25) = 8.25$

6. $23.89 \div (-3.48) = -6.8649425\ldots$ or -6.865 (rounded)

7. $3^2 + |-28| - (6^2 - 8) = 9 + 28 - (36 - 8) = 9 + 28 - 28 = 9$

8. $-\dfrac{2}{3} \div \dfrac{4}{5} - \left(-\dfrac{1}{2}\right) = -\dfrac{2}{3} \cdot \dfrac{5}{4} + \dfrac{1}{2} = -\dfrac{5}{6} + \dfrac{3}{6} = -\dfrac{2}{6} = -\dfrac{1}{3}$

9. 42,500

10. 510

11. 0.0000175

12. 0.61

13. 1.508×10^6

14. 8.5×10^0

15. 2.78×10^{-1}

16. 1.08×10^{-4}

17.
$$2x - 9 = -5x + 7$$
$$2x - 9 + 5x = -5x + 7 + 5x$$
$$7x - 9 = 7$$
$$7x - 9 + 9 = 7 + 9$$
$$7x = 16$$
$$x = \dfrac{16}{7} = 2\dfrac{2}{7}$$

18.
$$2x - 7x = 15$$
$$-5x = 15$$
$$x = -3$$

19.
$$3 = \dfrac{1}{3} - 2x$$
$$3(3) = 3\left(\dfrac{1}{3}\right) - 3(2x)$$
$$9 = 1 - 6x$$
$$9 - 1 = 1 - 6x - 1$$
$$8 = -6x$$
$$\dfrac{8}{-6} = -\dfrac{4}{3} = x$$

20.　　　　$-5(n - 2) = 8 - 4n$

$-5n + 10 = 8 - 4n$

$-5n + 10 + 4n = 8 - 4n + 4n$

$-n + 10 = 8$

$-n + 10 - 10 = 8 - 10$

$-n = -2$

$n = 2$

21.　　　　$\dfrac{2x}{3} - 5 = 7$

$3\left(\dfrac{2x}{3}\right) - 3(5) = 3(7)$

$2x - 15 = 21$

$2x - 15 + 15 = 21 + 15$

$2x = 36$

$x = 18$

22. $2x + 3(x - 5) = 15$

$2x + 3x - 15 = 15$

$5x - 15 = 15$

$5x - 15 + 15 = 15 + 15$

$5x = 30$

$x = 6$

23. $-2(x - 1) - 3x = 8 - 4x$

$-2x + 2 - 3x = 8 - 4x$

$-5x + 2 + 4x = 8 - 4x + 4x$

$-x + 2 - 2 = 8 - 2$

$-x = 6$

$x = -6$

24. $7(x - 2) - 6(x + 1) = -20$

$7x - 14 - 6x - 6 = -20$

$x - 20 = -20$

$x - 20 + 20 = -20 + 20$

$x = 0$

25. 12.5% of $150 = 0.125(150) = 18.75$

26. $x(80) = 10$

$x = \dfrac{10}{80} = 0.125$

$0.125 \times 100 = 12.5$ so $x = 12.5\%$

10

27. $4\%(x) = 64$

$0.04x = 64$

$x = \dfrac{64}{0.04} = 1600$

28. $I = Prt$. Substitute: $I = (\$2000)(0.08)(2) = \$320$

29. Let $\dfrac{76.8\,lb\ copper}{160\,lb\ alloy} = \dfrac{x}{100}$. Cross multiply.

$7680 = 160x$

$x = 48$ so there is 48% copper

30. $(1.5\%)(\$1800) = 0.015(1800) = \$27$ raise; New salary $= \$1800 + \$27 = \$1827$

31. Let $\dfrac{80}{100} = \dfrac{x}{2.00\,kg}$. Cross multiply.

$160 = 100x$

$1.6\ \text{kg} = x$

32. Calculate the expansion by multiplying $0.45\%(1.5671)$ and adding to the original length. $(0.45\%)(1.5671) = (0.0045)(1.5671) = 0.00705195$. Add this to 1.5671 to get an expanded length of 1.57415195 or 1.5742 meters (rounded).

33. $\dfrac{50 - 46.1}{50} \cdot 100 = \dfrac{3.9}{50} \cdot 100 = 7.8$ *so* 7.8% decrease in volume

34. $\dfrac{0.81 - 0.75}{0.75} \cdot 100 = \dfrac{0.06}{0.75} \cdot 100 = 8$ *so* 8% price increase

35. $\dfrac{\$5 - \$4}{\$4} \cdot 100 = \dfrac{1}{4} \cdot 100 = 25$, so 25% increase; Applying this to a \$9 former charge for a suit gives $\$9 + 0.25(\$9) = \$9 + \$2.25 = \$11.25$.

36. Let x = the original price of the ring. Savings = 40% of original price.

$x - 0.40x = \$599.99$

$0.60x = \$599.99$

$x = \$999.98$

37. Let x = her original hourly wage. Increase = 10% of original hourly wage.

$x + 0.10x = \$8.25$

$1.10x = \$8.25$

$x = \$7.50$

38. Let x = your friend's former weight. Loss = 12% of her former weight.

$$x - 0.12x = 110$$
$$0.88x = 110$$
$$x = 125 \text{ lb.}$$

39. 70in + 5% of 70in =70in + (0.05)(70in) = 70in + 3.5in = 73.5in = 6ft 1½in
    70in - 5% of 70in =70in - (0.05)(70in) = 70in - 3.5in = 66.5in = 5ft 6½in

40. total meal cost = $62.40 + 15% of $62.40 tip + 7% of $62.40 tax
    Total cost = $62.40 + (0.15)($62.40) + (0.07)($62.40)
    Total cost = $62.40 + $9.36 + $4.37 = $76.13
    Your part of the bill is one-third of $76.13 = ⅓($76.13) = $25.37666…=$25.38

Chapter R Test

1. The absolute value of a number is its distance from zero on the number line, and distances are always positive. Both 8 and -8 are eight units from zero on the number line.

2. It is the "same" number with the opposite sign or, it is the number with the same absolute value but on the opposite side of zero on the number line.

3. $2(5 - 7)^3 = 2(-2)^3 = 2(-8) = -16$

4. $|3^2 - 2(6 - 5)| = |3^2 - 2(1)| = |9 - 2| = 7$

5. $\dfrac{6(-5)}{2} + 3(6 - 9) - 7 = \dfrac{-30}{2} + 3(-3) - 7 = -15 + (-9) - 7 = -31$

6. 0.000208

7. 6.12×10^9

8.
$$3(y + 7) = 2y - 5$$
$$3y + 21 = 2y - 5$$
$$3y + 21 - 2y = 2y - 5 - 2y$$
$$y + 21 = -5$$
$$y + 21 - 21 = -5 - 21$$
$$y = -26$$

9.
$$\frac{5x}{6} = 2x - 7$$
$$6\left(\frac{5x}{6}\right) = 6(2x) - 6(7)$$
$$5x = 12x - 42$$
$$5x - 12x = 12x - 42 - 12x$$
$$-7x = -42$$
$$x = 6$$

10. $6x + x - 0.9 = 6x + 0.9$

$\quad\quad 7x - 0.9 = 6x + 0.9$

$\quad 7x - 0.9 - 6x = 6x + 0.9 - 6x$

$\quad\quad\quad x - 0.9 = 0.9$

$\quad x - 0.9 + 0.9 = 0.9 + 0.9$

$\quad\quad\quad\quad\quad x = 1.8$

11. $4y - 6(y + 4) = 1 - y$

$\quad 4y - 6y - 24 = 1 - y$

$\quad\quad\quad -2y - 24 = 1 - y$

$\quad -2y - 24 + y = 1 - y + y$

$\quad\quad\quad\quad -y - 24 = 1$

$\quad -y - 24 + 24 = 1 + 24$

$\quad\quad\quad\quad\quad\quad -y = 25$

$\quad\quad\quad\quad\quad\quad\quad y = -25$

12. $7 - (3x - 1) = 5x$

$\quad 7 - 3x + 1 = 5x$

$\quad\quad\quad 8 - 3x = 5x$

$\quad 8 - 3x + 3x = 5x + 3x$

$\quad\quad\quad\quad\quad 8 = 8x$

$\quad\quad\quad\quad\quad 1 = x$

13. 15% of $85 = 0.15 \times 85 = 12.75$

14. Using proportions: $\dfrac{x}{100} = \dfrac{33.5}{150}$

$\quad\quad\quad\quad\quad\quad 150x = 3350$

$\quad\quad\quad\quad\quad\quad\quad x = \dfrac{3350}{150} = 22.3\overline{3}$ *so approximately* 22.3%

15. Using proportions: $\dfrac{2.5}{100} = \dfrac{16.25}{x}$

$\quad\quad\quad\quad\quad\quad 2.5x = 1625$

$\quad\quad\quad\quad\quad\quad\quad x = 650$

16. Let x = sales tax.

$\quad$ Then $x = (6\%)(\$29.95) = (0.06)(29.95) = \$1.80$

17. Let x = tip. If 15% of *the cost of the meal* is the *tip*, then

$\quad\quad (0.15)(\$45.80) = x$

$\quad\quad\quad\quad\quad\$6.87 = x$

18. $\%$ decrease $= \dfrac{\$6500 - \$15{,}000}{\$15{,}000} = \dfrac{-\$8500}{\$15{,}000} = -0.56666 = -56.7\%$ or 57% decrease

19. If the voltmeter reads 6% too high, then the reading is 106% of the true voltage.
Let x = true voltage. Then, $106\%x = 120$

$$x = \frac{120}{106\%} = \frac{120}{1.06} = 113 \text{ volts (approximately)}$$

20. The error range is $\pm 0.15\%$(measurement) so

$(\pm 0.15\%)(0.453) = (\pm.0015)(0.453) = 0.001$

The upper boundary would be $0.453 + 0.001 = 0.454$

The lower boundary would be $0.453 - 0.001 = 0.452$

Solutions to Practice Problems in the Text

Chapter 1: Fundamentals of Mathematical Modeling

Practice Set 1-2

1. $d = rt$

    $125 = 50t$ [Divide both sides by 50.]

    $t = 2.5$ hours

3. $I = Prt$

    $I = (\$5000)(0.05)(2) = \$500$

5. $A = \pi r^2$

    $A = \pi(3)^2 = (3.14)(9) = 28.26 \text{ in}^2$

7. $A = \frac{1}{2}bh$

    $36 = \frac{1}{2}(8)h$

    $36 = 4h$

    $9 \text{ in} = h$

9. $A = \sqrt{s(s-a)(s-b)(s-c)} = \sqrt{15(15-5)(15-12)(15-13)} =$

    $\sqrt{15(10)(3)(2)} = \sqrt{900} = 30 \text{ in}^2$

11. $C = \frac{5}{9}(F-32) = \frac{5}{9}(68-32) = \frac{5}{9}(36) = 20^o \ C$

13. $F = \frac{9}{5}C + 32 = \frac{9}{5}(-10) + 32 = -18 + 32 = 14^o \ F$

15. $m = \dfrac{y_2 - y_1}{x_2 - x_1} = \dfrac{-3-(-4)}{-1-2} = \dfrac{-3+4}{-1-2} = \dfrac{1}{-3} = -\dfrac{1}{3}$

17. $z = \dfrac{x - \bar{x}}{s} = \dfrac{95-100}{15} = \dfrac{-5}{15} = -\dfrac{1}{3}$

19. $a^2 + b^2 = c^2$

    $3^2 + 4^2 = c^2$

    $9 + 16 = c^2$

    $25 = c^2$

    $\sqrt{25} = 5 = c$

21. $M = P\left(1 + \dfrac{r}{n}\right)^{nt} = 5000\left(1 + \dfrac{0.045}{12}\right)^{12*10} = 5000\left(1 + \dfrac{0.045}{12}\right)^{120} = \$7834.96$

23. $x = \dfrac{-b \pm \sqrt{b^2 - 4ac}}{2a} = \dfrac{-5 \pm \sqrt{(5)^2 - 4(1)(-6)}}{2(1)} = \dfrac{-5 \pm \sqrt{25 + 24}}{2} =$

$\dfrac{-5 + \sqrt{49}}{2} = \dfrac{-5 + 7}{2} = \dfrac{2}{2} = 1 \quad or \quad \dfrac{-5 - \sqrt{49}}{2} = \dfrac{-5 - 7}{2} = \dfrac{-12}{2} = -6$

25. $y = Ae^{rn} = 1{,}500{,}000e^{0.055*7} = 1{,}500{,}000e^{0.385} = 2{,}204{,}421$ bacteria

27. $I = Prt$

$\dfrac{I}{Pt} = \dfrac{Prt}{Pt}$ [Divide both sides by Pt.]

$\dfrac{I}{Pt} = r$

29. $A = \dfrac{1}{2}bh$

$2A = bh$ [Multiply both sides by 2.]

$\dfrac{2A}{h} = b$ [Divide both sides by h to solve for b.]

31. $P = 2L + 2W$

$P - 2W = 2L + 2W - 2W$ [Subtract $2W$ from both sides.]

$P - 2W = 2L$

$\dfrac{P - 2W}{2} = L$ [Divide both sides by 2.]

33. $A = \frac{1}{2}(B + b)h$

$2(A) = 2[\frac{1}{2}(B + b)h]$ [Multiply both sides by 2.]

$2A = (B + b)h$

16

$$2A = Bh + bh \qquad \text{[Use the Distributive Property.]}$$

$$2A - bh = Bh + bh - bh \qquad \text{[Subtract } bh \text{ from both sides.]}$$

$$2A - bh = Bh$$

$$\frac{2A - bh}{h} = B \qquad \text{[Divide both sides by } h \text{ to solve for } B.]$$

35. $$2x + 3y = 6$$

$$2x + 3y - 2x = 6 - 2x \qquad \text{[Subtract } 2x \text{ from both sides.]}$$

$$3y = 6 - 2x$$

$$y = 2 - \tfrac{2}{3}x \qquad \text{[Divide both sides by 3 to solve for } y.]$$

37. $$A = \frac{x+y}{2}$$

$$2A = x + y \qquad \text{[Multiply both sides by 2.]}$$

$$2A - y = x + y - y \qquad \text{[Subtract } y \text{ from both sides.]}$$

$$2A - y = x$$

39. $$F = \frac{9}{5}C + 32$$

$$5F = 9C + 160 \qquad \text{[Multiply through by 5.]}$$

$$5F - 160 = 9C + 160 - 160 \qquad \text{[Subtract 160 from both sides.]}$$

$$5F - 160 = 9C$$

$$\frac{5}{9}F - \frac{160}{9} = C \qquad \text{[Divide both sides by 9.]}$$

$$\frac{5}{9}(F - 32) = C \qquad \text{[Factor out } \frac{5}{9}.]$$

41. $$\text{BMI} = \frac{W}{H^2} \cdot 703 = \frac{140}{73^2} \cdot 703 = 18.4687... = 18.5 \text{, Normal}$$

Practice Set 1-3

1. $\dfrac{45\,min}{2\,hr} = \dfrac{45\,min}{120\,min} = \dfrac{3}{8}$

3. $\dfrac{4\,in}{4\,ft} = \dfrac{4\,in}{48\,in} = \dfrac{1}{12}$

5. $\dfrac{6\,ft}{3\,yd} = \dfrac{6\,ft}{9\,ft} = \dfrac{2}{3}$

7. $\dfrac{8\,weeks}{16\,days} = \dfrac{56\,days}{16\,days} = \dfrac{7}{2}$

9. $\dfrac{25\,mL}{1\,L} = \dfrac{25\,mL}{1000\,mL} = \dfrac{1}{40}$

11. $\dfrac{304\,miles}{9.5\,gal} = 32$ mi./gal.

13. $\dfrac{\$3.50}{10\,min} = \$0.35/min.$

15. $\dfrac{\$48}{10\,days} = \$4.80/day$

17. $\dfrac{24\,lb}{15\,people} = 1.6$ lb./person

19. $\dfrac{x}{5} = \dfrac{3}{4}$

    $4x = (3)(5)$ [cross-multiplication property]

    $4x = 15$

    $x = \dfrac{15}{4}$

21. $\dfrac{30}{126} = \dfrac{5}{3x}$

    $(3x)(30) = 5(126)$ [cross-multiplication property]

    $90x = 630$

    $x = 7$

23. $\dfrac{3x+6}{35} = \dfrac{2x-18}{5}$

    $5(3x + 6) = 35(2x - 18)$ [cross-multiplication property]

    $15x + 30 = 70x - 630$ [distributive property]

$$15x + 30 - 70x = 70x - 630 - 70x$$
$$-55x + 30 = -630$$
$$-55x + 30 - 30 = -630 - 30$$
$$-55x = -660$$
$$x = 12$$

25. $\dfrac{15}{18} = \dfrac{x-1}{x}$

         $(15)(x) = 18(x - 1)$ [cross-multiplication property]

          $15x = 18x - 18$ [distributive property]

     $15x - 18x = 18x - 18 - 18x$

          $-3x = -18$

          $x = 6$

27. $\dfrac{3}{x+1} = \dfrac{18}{9x-3}$

         $3(9x - 3) = 18(x + 1)$ [cross-multiplication property]

         $27x - 9 = 18x + 18$ [distributive property]

    $27x - 9 - 18x = 18x + 18 - 18x$

         $9x - 9 = 18$

      $9x - 9 + 9 = 18 + 9$

          $9x = 27$

          $x = 3$

29.   Unit rate equals cost divided by the number of square feet.

      $\$2235 \div 1500 \text{ ft}^2 = \$1.49/\text{ft}^2$

31.   Cost per ounce = $\$8.99 \div 16$ oz = $\$0.561875$ per ounce

33.   Brand X: $\$1.49 \div 8$ oz = $\$0.18625/$ ounce

      Brand Z: $\$2.12 \div 12$ oz = $\$0.1766\ldots/$ounce

      Brand Z is the better buy because it costs less per ounce.

35.   Let $\dfrac{120 \ cal}{\frac{3}{4} \ cup} = \dfrac{x}{1 \ cup}$.

         $(1)(120) = \tfrac{3}{4}x$ [Cross multiply.]

          $120 = \tfrac{3}{4}x$

     160 calories $= x$ [Multiply both sides by 4/3.]

37.   Let $\dfrac{2.5 \ dozen}{1.25 \ cups} = \dfrac{x \ dozen}{3 \ cups}$.

         $(3)(2.5) = 1.25x$ [Cross multiply.]

          $7.5 = 1.25x$

$$6 = x$$

Answer is 6 dozen or 72 muffins.

39. Let $\dfrac{5\,ft\,4\,in}{10.5\,ft} = \dfrac{x}{20\,ft}$. Convert the measurements to inches as follows: 5 ft 4 in = 5(12) + 4 = 64 inches; 10.5 ft = 10.5 × 12 in = 126 inches; 20 ft = 20 × 12 in = 240 inches. Now substitute:

$$\dfrac{64\,in}{126\,in} = \dfrac{x}{240\,in}$$
$$(64)(240) = 126x$$
$$15360 = 126x$$
$$121.9 \text{ inches} = x$$

Converting to feet and inches: 121.9 inches ÷ 12in/ft = 10 ft. 1.9 in.

41. Let $\dfrac{1.8\,A}{18\,V} = \dfrac{5.4\,A}{x}$.

$$(18)(5.4) = 1.8A \qquad \text{[Cross multiply.]}$$
$$97.2 = 1.8A$$
$$54\,V = A$$

43. Let $\dfrac{110\,lb}{19.4\,lb} = \dfrac{200\,lb}{x}$

$$110x = (19.4)(200) \qquad \text{[Cross multiply.]}$$
$$110x = 3880$$
$$x = 35.2727\ldots \text{ or approximately } 35.3 \text{ lb.}$$

45. Let $\dfrac{10\,lb}{400\,ft^2} = \dfrac{x}{500\,ft^2}$

$$(10)(500) = 400x \qquad \text{[Cross multiply.]}$$
$$5000 = 400x$$
$$12.5 \text{ lb} = x$$

47. Let $\dfrac{\$240}{1200\,words} = \dfrac{x}{1500\,words}$.

$$(240)(1500) = 1200x \qquad \text{[Cross multiply.]}$$
$$360,000 = 1200x$$
$$\$300 = x$$

49. Let $\dfrac{1\,adult}{15\,children} = \dfrac{3\,adults}{x}$

$$x = (3)(15) \qquad \text{[Cross multiply.]}$$
$$x = 45 \text{ children}$$

51. Let $\dfrac{1\ inch}{8\ feet} = \dfrac{2.75\ inches}{x}$

$\quad\quad\quad\quad\quad\quad x = (8)(2.75)$ [Cross multiply.]

$\quad\quad\quad\quad\quad\quad x = 22$ feet

Let $\dfrac{1\ inch}{8\ feet} = \dfrac{1.9375\ inches}{x}$

$\quad\quad\quad\quad\quad\quad x = (8)(1.9375)$ [Cross multiply.]

$\quad\quad\quad\quad\quad\quad x = 15.5$ feet

Dimensions are 22 ft × 15.5 ft

53. Let $\dfrac{Mach\ 1}{761.2\ mph} = \dfrac{Mach\ 3.1}{x}$

$\quad\quad\quad\quad\quad 1x = (3.1)(761.2)$ [Cross multiply.]

$\quad\quad\quad\quad\quad x = 2359.72$ mph

55. Let $\dfrac{400\ ft}{1\ day} = \dfrac{1000\ ft}{x}$

$\quad\quad\quad (1)(1000) = 400x$ [Cross multiply.]

$\quad\quad\quad\quad\ 1000 = 400x$

$\quad\quad\ 2.5\ days = x$

57. Let $\dfrac{40\ mg}{1\ mL} = \dfrac{60\ mg}{x}$.

$\quad\quad\quad\quad 40x = (1)(60)$ [Cross multiply.]

$\quad\quad\quad\quad 40x = 60$

$\quad\quad\quad\quad\ \ x = 1.5$ mL

59. Let $\dfrac{1.8\ mi}{30\ min} = \dfrac{x}{45\ min}$

$\quad\quad\ (1.8)(45) = 30x$ [Cross multiply.]

$\quad\quad\quad\quad\ \ 81 = 30x$

$\quad\ 2.7\ miles = x$

Practice Set 1-4

1. $2x + 6$

3. $7x - 2$

5. $3(4 + x)$

7. $\dfrac{1}{3}x - 5$

9. Let Bill's salary = x.   Then, Ann's salary = x + $5000

11. Let the width = x. Then, the length = 2x + 5.

13.
$$5x + 5 = 2x - 10$$
$$5x + 5 - 2x = 2x - 10 - 2x$$
$$3x + 5 = -10$$
$$3x + 5 - 5 = -10 - 5$$
$$3x = -15$$
$$x = -5$$

15. Emily has saved x. Elena has saved 2x (twice as much).
$$x + 2x = 72$$
$$3x = 72$$
$$x = 24 \text{ so Emily has saved } \$24 \text{ and Elena has saved } 2(\$24) = \$48.$$

17. Let x = the shorter piece. The longer piece will be 2x.
$$x + 2x = 60 \text{ meters}$$
$$3x = 60 \text{ meters}$$
$$x = 20 \text{ meters}$$
The shorter piece is 20 meters long and the longer piece is 2(20) = 40 meters long.

19. Using the definition of an average, let x equal the missing grade.
$$\frac{88 + 91 + 95 + x}{4} = 90$$
$$88 + 91 + 95 + x = 4(90)$$
$$274 + x = 360$$
$$274 - 274 + x = 360 - 274$$
$$x = 86$$
In order to have a 90 average, she must make 86 on the last test.

21. Let x = the number of kilowatt-hours. Write the equation and solve.
$$\$20.00 + \$0.14x = \$85.78$$
$$20.00 - 20.00 + 0.14x = 85.78 - 20.00$$
$$0.14x = 65.78$$
$$x = \frac{65.78}{0.14} = 470 \, kWh \quad \text{(rounded)}$$

23. Let x = the taxi fare. Since 1/11 of a mile costs $0.20, the cost of one mile = $0.20 × 11 = $2.20.
$$x = \$3.20 + 18(\$2.20/mile) + \$1.20 = \$44.00$$

25. The opponent scored x points in the game. Mighty Mites scored 39 points. This amount (39) is 2x - 1.
$$2x - 1 = 39$$
$$2x - 1 + 1 = 39 + 1$$
$$2x = 40$$
$$x = 20$$
The opponent scored 20 points.

27. Let the integers be x, x + 1 and x + 2. Write the equation for the sum and solve.

$$x + (x + 1) + (x + 2) = 87$$
$$3x + 3 = 87$$
$$3x + 3 - 3 = 87 - 3$$
$$3x = 84$$
$$x = 28$$

The integers are x = 28, x + 1 = 29 and x + 2 = 30.

29. Let the odd integers be x, x + 2, and x + 4. Write the equation for the sum and solve.

$$x + (x + 2) + (x + 4) = -273$$
$$3x + 6 = -273$$
$$3x + 6 - 6 = -273 - 6$$
$$3x = -279$$
$$x = -93$$

The three integers are x = -93, x + 2 = -91, and x + 4 = -89.

31. Let x = the value of the lot. Then, 6.5x = the value of the house.

$$x + 6.5x = \$175{,}000$$
$$7.5x = \$175{,}000$$
$$x = \$23{,}333.33$$

So, the lot is worth approximately \$23,333 and the house is worth about \$151,667.

33. $x = \dfrac{949 \ yen}{146 \ yen / dollar} = \$6.50$.

35. $time = \dfrac{distance}{speed} = \dfrac{500mi}{170.265mi / hr} = 2.936598831hr = $ about 2 hr 56 min

37. Let the number of males and females at the beginning of the semester = x. If 8 males drop the class, there are x − 8 males remaining. Now the number of females is twice the number of males remaining.

$$x = 2(x - 8)$$
$$x = 2x - 16$$
$$x - 2x = 2x - 16 - 2x$$
$$-x = -16$$
$$x = 16$$

There were 16 males and 16 females at the beginning of the semester.

39. Let x = the number of bags of apples.

$$5x + 2x = 252$$
$$7x = 252$$
$$x = 36$$

There are 36 bags containing 5 lb. of apples and 36 bags containing 2 lb. of apples.

41. Let x = profit from the automotive division. Then x – 273 represents the profit from financial services.

$$x + (x - 273 \text{ million}) = 483 \text{ million}$$
$$2x - 273 \text{ million} = 483 \text{ million}$$
$$2x - 273 \text{ million} + 273 \text{ million} = 483 \text{ million} + 273 \text{ million}$$
$$2x = 756 \text{ million}$$
$$x = 378 \text{ million}$$

The profit from the automotive division was 378 million and from financial services was 378 million – 273 million = 105 million.

43. Let x = attendance at the Ohio State game and x – 1044 = attendance at the Penn State game. Total attendance for the two teams in 2010 was 209,512.

$$x + x - 1044 = 209,512$$
$$2x - 1044 = 209,512$$
$$2x - 1044 + 1044 = 209,512 + 1044$$
$$2x = 210,556$$
$$x = 105,278$$

Therefore, the attendance at the Ohio State game was 105,278 and the attendance at the Penn State game was 105,278 – 1044 = 104,234.

45. Let x = the original price of the radio. Savings = 15% of retail price or 0.15x.

$$x - 0.15x = \$127.46$$
$$0.85x = \$127.46$$
$$x = \$149.95$$

The original price of the radio was $149.95.

47. Let x = the wholesale price of the shoes. The markup amount is 65% times x.

$$x + 0.65x = \$125.40$$
$$1.65x = \$125.40$$
$$x = \$76$$

The wholesale price of the shoes was $76.00.

49. Let x = Drema's contributions. The company's contributions = 20% of x.

$$x + 0.20x = \$1200$$
$$1.2x = \$1200$$
$$x = \$1000$$

Drema deposited $1000 into the account.

51. $\text{BMI} = \dfrac{W}{H^2} \cdot 703 = \dfrac{120}{60^2} \cdot 703 = 23.4333... = 23.4$, Normal

Chapter 1 Review Problems

1.. $I = Prt$

$t = \dfrac{I}{Pr}$ [Divide both sides by Pr.]

2. $2x + 3y = 9$

$3y = -2x + 9$ [Subtract $2x$ from both sides.]

$y = -\dfrac{2}{3}x + 3$ [Divide all terms by 3.]

3. $C = \pi d$

$d = \dfrac{C}{\pi}$ [Divide both sides by π.]

4. $a + b + c = P$

$c = P - a - b$ [Subtract a and b from both sides.]

5. $\dfrac{27\,min}{3\,hr} = \dfrac{27\,min}{180\,min} = \dfrac{3}{20}$

6. $\dfrac{4\,weeks}{21\,days} = \dfrac{28\,days}{21\,days} = \dfrac{4}{3}$

7. $\dfrac{6\,in}{5\,ft} = \dfrac{6\,in}{60\,in} = \dfrac{1}{10}$

8. $\dfrac{\$60}{5\,hr} = \$12/\text{hour}$

9. $\dfrac{44\,bushels}{8\,trees} = 5.5$ bushels/tree

10. $\dfrac{\$12.80}{3.5\,lb} = \$3.66/\text{lb.}$

11. $\dfrac{x}{3} = \dfrac{4}{7}$

$7x = 12$

$x = \dfrac{12}{7}$

12. $\dfrac{2}{3} = \dfrac{8}{2x}$

$4x = 24$

$x = 6$

13. $\dfrac{x-3}{8} = \dfrac{3}{4}$

$4x - 12 = 24$

$4x - 12 + 12 = 24 + 12$

$4x = 36$

$x = 9$

14. $\dfrac{4x-3}{7} = \dfrac{2x-1}{3}$

$12x - 9 = 14x - 7$

$12x - 9 - 14x = 14x - 7 - 14x$

$-2x - 9 + 9 = -7 + 9$

$-2x = 2$

$x = -1$

15. $A = \frac{1}{2}bh$. Substitute: $A = \frac{1}{2}(3 \text{ in})(4 \text{ in}) = 6 \text{ in}^2$

16. Let $\dfrac{2400\,L}{50\,min} = \dfrac{x}{30\,min}$. Cross multiply.

$72000 = 50x$

$1440 \text{ liters} = x$

17. Let $\dfrac{3}{4} = \dfrac{x}{92}$. Cross multiply.

$276 = 4x$

$69 \text{ dentists} = x$

18. Let x = total number of minutes that you talk. Then, $\$1.24 + (x - 4)(\$0.28) = \$3.76$

$1.24 + 0.28x - 1.12 = 3.76$

$0.12 + 0.28x = 3.76$

$0.12 + 0.28x - 0.12 = 3.76 - 0.12$

$0.28x = 3.64$

$x = 13$ The call was 13 minutes long.

19. Let x = number of minutes per month. Then, $\$50 + \$0.36x = \$99.68$.

$0.36x = 99.68 - 50$

$0.36x = 49.68$

$x = 138 \text{ minutes}$

26

20. Let x = labor time (per hour). Then, $40 + $30x = $115

$$40 + 30x - 40 = 115 - 40$$
$$30x = 75$$
$$x = 2.5 \text{ hours}$$

21. Let the shorter piece be *x* and the longer be *x + 3*. Then, x + (x + 3) = 33

$$2x + 3 = 33$$
$$2x + 3 - 3 = 33 - 3$$
$$2x = 30$$
$$x = 15 \text{ in (short piece)} \quad \text{and} \quad x + 3 = 15 + 3 = 18 \text{ in (long piece)}$$

22. Let the short piece = *x*, the middle-sized piece = *x + 14,* and the long piece = *3x.*

Then, x + (x + 14) + 3x = 79
$$5x + 14 = 79$$
$$5x + 14 - 14 = 79 - 14$$
$$5x = 65$$
$$x = 13 \text{ cm (short)}$$
$$x + 14 = (13) + 14 = 27 \text{ cm (middle)}$$
$$3x = 3(13) = 39 \text{ cm (long)}$$

23. Let the calculator be *x* and the cassette player be *x + 140.*

Then, x + (x + 140) = 208
$$2x + 140 = 208$$
$$2x + 140 - 140 = 208 - 140$$
$$2x = 68$$
$$x = \$34.00 \text{ which is the cost of the calculator}$$

24. Let *n* = the number of nickels and *2n - 2* = the number of dimes.

Then, n + (2n - 2) = 52.
$$3n - 2 = 52$$
$$3n - 2 + 2 = 52 + 2$$
$$3n = 54$$
$$n = 18 \quad \text{There are 18 nickels and 34 dimes in the bank for a total of \$4.30.}$$

25. Use the formula *d = rt* and substitute values.

$$182 \text{ mi} = (52 \text{ mph})(t)$$
$$t = 3.5 \text{ hours}$$

26. *T= UN + F*   Substitute the given values and solve.

$$\$16{,}750 = \$15N + \$2500$$
$$16750 - 2500 = 15N + 2500 - 2500$$
$$14250 = 15N$$
$$N = 950 \text{ units produced}$$

26. First remove the parentheses by use of the distributive property, then add 2 to the 12.

$$2 + 3(2x + 4) = 2 + 6x + 12 = 6x + 14.$$

28.
$$5x + 3 = 6x$$
$$5x + 3 - 5x = 6x - 5x$$
$$3 = x$$

29.   for any triangle, $A + B + C = 180°$
Let A = the first angle
The second angle, $B = 3A$
The third angle, $C = \frac{2}{3}(3A) = 2A$
$A + 3A + 2A = 180°$
$6A = 180°$
$A = 30°, B = 3(30°) = 90°, C = 2(30°) = 60°$

30.   First convert all the measurements to the same units, like inches.
5 ft 3 in = 63 in,   10 ft 6 in = 126 in, and 52 ft = 624 in

Now set up a proportion like: $\dfrac{63in}{126in} = \dfrac{x}{624in}$

Cross multiply: $(63in)(624in) = (126in)(x)$
Divide by 126in: $x = 312in = 26ft$

Chapter 1 Test

1. $V = lwh$

    $w = \dfrac{V}{lh}$ [Divide both sides by *lh*.]

2. $h = vt - 16t^2$
    $h + 16t^2 = vt$ [Add *16t²* to both sides.]

    $\dfrac{h + 16t^2}{t} = v$ [Divide both sides by *t*.]

3. $\dfrac{4\,hr}{1\,day} = \dfrac{4\,hr}{24\,hr} = \dfrac{1}{6}$

4. $\dfrac{10\,ft}{160\,in} = \dfrac{120\,in}{160\,in} = \dfrac{3}{4}$

5. $\dfrac{\$413.20}{4\,days} = \$103.30/day$

28

6. $\dfrac{7.5 \ lb}{6 \ people} = 1.25 \ lb/person$

7. $\dfrac{50 \ eggs}{10 \ chickens} = 5 \ eggs/chicken$

8. $\dfrac{7}{12} = \dfrac{3x}{10}$

   $70 = 36x$

   $\dfrac{70}{36} = \dfrac{35}{18} = x$

9. $\dfrac{x-4}{8} = \dfrac{2x+3}{9}$

   $9(x - 4) = 8(2x + 3)$

   $9x - 36 = 16x + 24$

   $9x - 36 - 16x = 16x + 24 - 16x$

   $-7x - 36 = 24$

   $-7x - 36 + 36 = 24 + 36$

   $-7x = 60$

   $x = -\dfrac{60}{7}$

10. Let x = defective bulbs. Then, $\dfrac{3}{85} = \dfrac{x}{510}$.

    $85x = 1530$

    $x = 18$ bulbs

11. Let x = parts produced. Then, $\dfrac{300 \ parts}{20 \ min} = \dfrac{x}{45 \ min}$.

    $20x = 13500$

    $x = 675$ parts

12. Given $P = 2L + 2W$. Substitute the given values into the formula.

    $P = 2(20 \ ft.) + 2(12 \ ft.) = 64 \ ft.$

13. Company A Plan: $20 + 0.10m$ (where m = miles driven)

    Company B Plan: $10 + 0.30m$

    To find the number of miles where the two costs are equal, set these two expressions equal.

    $\$20 + 0.10m = \$10 + 0.30m$

    $20 + 0.10m - 0.10m = 10 + 0.30m - 0.10m$

    $20 = 10 + 0.20m$

    $20 - 10 = 10 + 0.20m - 10$

    $10 = 0.20m$

    $50 = m$, so at 50 miles the costs are the same for both plans

29

14. Let the number of passengers on one ship x.
    The second ship holds twice as many passengers $= 2x$
    Write the equation $x + 2x = 2250$ and solve.
    $$3x = 2250 \qquad =$$
    $$x = 750 \text{ so the smaller ship holds 750 passengers}$$

15. Let Sarah's age be x. Then, Michelle's age is *5x - 10*. Write the equation and solve.
    $$x + 5x - 10 = 44$$
    $$6x - 10 = 44$$
    $$6x - 10 + 10 = 44 + 10$$
    $$6x = 54$$
    $$x = 9 \quad \text{so Sarah is 9 years and Michelle is 5(9)-10 = 35 years.}$$

16. Let the short board $= x$ and the longer one be *3x + 1*. Write the equation.
    $$x + 3x + 1 = 21$$
    $$4x + 1 = 21$$
    $$4x + 1 - 1 = 21 - 1$$
    $$4x = 20$$
    $$x = 5 \quad \text{so the short board is 5 ft. and the longer is } 3(5) + 1 = 16 \text{ ft.}$$

17. $x + 12.3\%$ of x is now \$2.83 per gallon
    $x + 0.123x = \$2.83$
    $1.123x = \$2.83$
    $x = \$2.52$ per gallon

18. $P - 22\%$ of P is now 28,000 people, after 8 years
    $P - 0.22P = 28,000$
    $0.78P = 28,000$
    $P = 35,897.4359 = 35,898$ whole people 8 years ago
    Now divide the amount of decrease in population by 8 years to get a rough average yearly decrease in population:
    $35,898 - 28,000 = 7898$ people in 8 years
    $7898/8 = 987.25$ or about 988 people per year decrease in population

19. The average of the four grades needs to equal at least 84 so:
    $(75 + 82 + 80 + x)/4 = 84$ [multiply by 4]
    $75 + 82 + + 80 + x = 336$
    $237 + x = 336$
    $x = 99$ is the least the student can make an have an overall average of 84

20. Let L = lot value, H = house value = 7.5L

    L + 7.5L = $152,000

    8.5L = $152,000

    L = $17,882.35294 or about $17,882 for the lot

    H = 7.5L = $134,117.6471 or about $138,118 for the house

Chapter 2: Applications of Algebraic Modeling

Practice Set 2-1

1. Let L = 360 ft and W = 160 ft.

    P = 2L + 2W

    P = 2(360 ft) + 2(160 ft) = 720 ft + 320 ft = 1040 ft

3. Let L = 15 ft and W = 11 ft.

    A = LW = (15 ft)(11 ft) = 165 ft^2

    1 yd^2 = 9 ft^2 so divide by 9 to convert to square yards.

    165 ft^2 ÷ 9 ft^2/yd^2 = 18.3 yd^2

5. A square has 4 equal sides. P = 4L = 54 in., so L = 13.5 in per side. The area of a square is given by the formula: A = L^2 = (13.5 in)^2 = 182.25 in^2.

7. Use the formula for the area of a rectangle to calculate the size of the area to be planted.

    A = lw = (8.75 ft)(4.5 ft) = 39.375 ft^2

    Since she needs four plants for every square foot, you should multiply this answer by 4. This gives 39.375 * 4 = 157.5 plants. In a practical sense, using this answer means you should buy 158 plants. If you round the area off to 40 ft^2 and multiply by 4, you would get 160 plants.

9. A = lw = (220ft)(158ft) = 34,760 ft^2

    Calculate the number of bags needed.

    34,760 ft^2 ÷ 1000 ft^2/bag = 34.76 = 35 bags of grass seed

11. The label is a rectangle with one side being the height of the can and the other the circumference of the circular end of the can.

    A = lw = (4.5 in.)(7.9 in.) = 35.6 in.^{2} of paper

13. The paper is rectangular so it will cover an area of A = lw = (20 in)(15 in) = 300 in^2.

    The top, sides and bottom of the box must be covered so we will calculate the area of each side of the box and total the area.

$$\text{Top} = \text{lw} = (12 \text{ in.})(8 \text{ in.}) = 96 \text{ in}^2$$

$$\text{Side} = \text{lw} = (12 \text{ in.})(3 \text{ in.}) = 36 \text{ in}^2$$

$$\text{End} = \text{lw} = (8 \text{ in.})(3 \text{ in.}) = 24 \text{ in}^2$$

For the whole box, there are six surfaces, (top/bottom, 2 ends, 2 sides) so the total area is $2(96 \text{ in}^2) + 2(36 \text{ in}^2) + 2(24 \text{ in}^2) = 312 \text{ in}^2$. Therefore, this piece of paper is not big enough to cover the box.

15. Divide the areas by the cost for each pizza to find the least costly pizza. Use the formula $A = \pi r^2$ for each pizza.

8 in. diameter pizza: $A = \pi(4 \text{ in.})^2 = 50.265 \text{ in.}^2 \div 699 \text{ cents} = 0.0715 \text{ in}^2/\text{cent}$

16 in. diameter pizza: $A = \pi(8 \text{ in.})^2 = 201.062 \text{ in.}^2 \div 1595 \text{ cents} = 0.126 \text{ in}^2/\text{cent}$

So, you get more square inches of pizza to eat per cent with the 16-inch pizza.

17. Find the area of the square. Then find the area of the circle and subtract the areas.

$$\text{Area of the square} = s^2 = (24 \text{ m})^2 = 576 \text{ m}^2$$

$$\text{Area of the circle} = \pi r^2 = \pi \cdot 12^2 = \pi \cdot 144 \approx 452.4 \text{ m}^2$$

$$\text{Area of the shaded part} = 576 \text{ m}^2 - 452.4 \text{ m}^2 = 123.6 \text{ m}^2$$

19 a) cost divided by the area $= \dfrac{\$8625}{(50\,ft)(230\,ft)} = \dfrac{\$8625}{11500\,ft^2} = \$0.75/ft^2$

b) $\$8625/50$ ft $= \$172.50$ per foot of frontage

c) $(43560 \text{ ft}^2/\text{acre})(\$0.75/\text{ft}^2) = \$32,670/\text{acre}$

21. $A = l^2$ and $P = 4l$

a) $A = (2l)^2 = 4l^2$ or 4(original area) and $P = 4(2l) = 8l$ or 2 × original perimeter

b) $A = (0.5l)^2 = 0.25l^2$ or ¼(original area) and $P = 4(0.5l) = 2l$ or ½ of original perimeter

c) $A = (3l)^2 = 9l^2$ or 9(original area) and $P = 4(3l) = 12l$ or 3 × original perimeter

d) Doubling the length of the diagonal of a square will double the side lengths of the square thus, the results will be the same as in (a) above.

23. The line formed by the diameters of four circles in a row with their sides touching each other and the sides of the square will equal the length of the rectangle and the line formed by two diameters of circles in a column will equal the width of the rectangle.

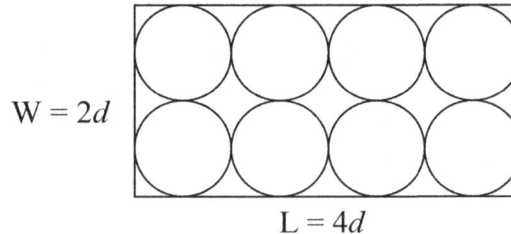

$$W = 2d$$

$$L = 4d$$

Find the area of 8 equal circles based on a diameter of d for each ($r = 0.5\,d$) then divide that by the area of a rectangle with L = 4d and W = 2d.

$A_{\text{circle}} = \pi r^2 = \pi(0.5d)^2 = 0.25\pi d^2 = 0.785\ldots\, d^2$, so 8 circles have an area of about $6.283d^2$

$A_{\text{rectangle}} = LW = (4d)(2d) = 8d^2$

$$\frac{A_{circle}}{A_{rectangle}} = \frac{6.283d^2}{8d^2} = 0.785 = 78.5\%$$

25. a) Start by multiplying the formula given by 2 to clear the fraction.

$$2(w + 2h + \tfrac{\pi w}{2}) = 2(20 \text{ ft})$$

$$2w + 4h + \pi w = 40 \text{ ft}$$

$$4h = 40 \text{ ft} - 2w - \pi w$$

$$h = \frac{40\,ft}{4} - \frac{2w}{4} - \frac{\pi w}{4}$$

$$h = 10 \text{ ft} - 0.5w - 0.785w = 10 \text{ ft} - 1.285w \approx 10 \text{ ft} - 1.3w$$

b) Base the graph on the intercepts, h = 0 and w = 0

0 = 10 - 1.285w

w = 7.78, which gives one intercept, (7.78 , 0)

h = 10 - 1.285(0)

h = 10, which gives the other intercept, (0 , 10)

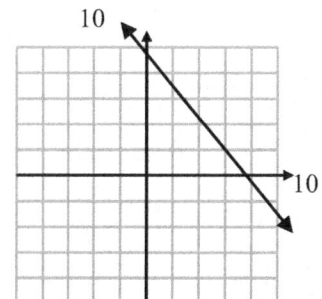

c) We chose for w to be simple lengths or 0, 1, 2, and 3 feet, then solved for h.

   for w = 0, h = 10 - 1.285w = 10 - 1.285(0) = 10

   for w = 1, h = 10 - 1.285(1) = 8.715

   for w = 2, h = 10 - 1.285(2) = 7.430

   for w = 3, h = 10 - 1.285(3) = 6.145

d) Theoretically one could choose any positive real number to represent the width of the window and this would lead to an infinite number of possibilities.

27. d = 8ft, so r = 4ft

$A = \pi r^2 = \pi 4^2 = 50.265... \approx 50.3$ ft

29. $A_{circle} - A_{pool} = \pi(6)^2 - \pi(4)^2 = 20\pi$ ft^2

(4 pavers per square foot)(20π ft^2) = 80π pavers needed

80π ($0.59) = $148.28

Practice Set 2-2

1. $180° - \angle A - \angle B = 180° - 35° - 55° = 90°$; right triangle

3. $180° - \angle A - \angle B = 180° - 101° - 36° = 43°$; obtuse triangle

5. $180° - \angle A - \angle B = 180° - 90° - 29° = 61°$; right triangle

7. $A = \frac{1}{2}bh = \frac{1}{2}(2.5 \text{ cm})(3 \text{ cm}) = 3.75$ cm^2

9. $A = \frac{1}{2}bh = \frac{1}{2}(5 \text{ in.})(12 \text{ in.}) = 30$ in^2

11. $A = \frac{1}{2}bh = \frac{1}{2}(175 \text{ mm})(100 \text{ mm}) = 8750$ mm^2

13. Use Hero's formula s $= \frac{15 + 15 + 20}{2} = 25$

   $A = \sqrt{25(25 - 15)(25 - 15)(25 - 20)} = \sqrt{25 \cdot 10 \cdot 10 \cdot 5} = \sqrt{12500} = 111.8$ in.^{2}

15. Area of Rectangle = lw = (74 cm)(48 cm) = 3552 cm^2

   Area of Triangle $= \frac{1}{2}bh = \frac{1}{2} \cdot 37\text{cm} \cdot 48 \text{ cm} = 888$ cm^2

   Shaded Portion = $A_R - A_T = 3552$ cm^2 – 888 cm^2 = 2664 cm^2

17. Use Hero's formula for triangle: s $= \frac{10 + 10 + 10}{2} = 15$

   $A = \sqrt{15(15 - 10)(15 - 10)(15 - 10)} = \sqrt{15 \cdot 5 \cdot 5 \cdot 5} = \sqrt{1875} \approx 43.30$ cm^2

36

Area of Circle $= \pi r^2 = \pi(3 \text{ cm})^2 \approx 28.27 \text{ cm}^2$

Shaded Portion $= A_T - A_C = 43.30 \text{ cm}^2 - 28.27 \text{ cm}^2 = 15.03 \text{ cm}^2$

19. A right triangle has one 90° angle and two acute angles.

    The total degree measure is 180°.

    Third angle = 180° - 90° - 45° = 45°. Because the two acute angles are equal, this triangle is isosceles.

21. An isosceles triangle has two angles that are equal. The total degree measure is 180°.

    Third angle = 180° - 32° - 32° = 116°. Because this angle is > 90°, the triangle is an obtuse triangle.

23. Area of Triangle $= \dfrac{1}{2} bh$

    $132 \text{ in.}^2 = \dfrac{1}{2}(24 \text{ in.})h$

    $132 = 12 h$

    $11 \text{ in.} = h$

25. Hero's formula: $s = \dfrac{18+11+25}{2} = 27$

    $A = \sqrt{27(27-18)(27-11)(27-25)} = \sqrt{7776} = 88.18 \text{ ft}^2$

27. Use Hero's formula: $s = \dfrac{10.5 + 8.25 + 5.25}{2} = 12$

    $A = \sqrt{12(12-10.5)(12-8.25)(12-5.25)} = \sqrt{12(1.5)(3.75)(6.75)} =$

    $\sqrt{455.625} = 21.3 \text{ ft}^2$

29. Use Hero's formula: $s = \dfrac{756 + 612 + 612}{2} = 990$

    $A = \sqrt{990(990-756)(990-612)(990-612)} = \sqrt{990(234)(378)(378)} =$

    $\approx 181{,}935 \text{ ft}^2$

Practice Set 2-3

1. The triangle must be a right triangle, and you must know the length of two sides.

37

Note: for problems 3 – 7 the Pythagorean Theorem is used with 'c' being the hypotenuse.

3. $c^2 = 9^2 + 12^2 = 81 + 144 = 225$

    $c = \sqrt{225} = 15.0$

5. $12^2 = a^2 + 6^2$; $12^2 - 6^2 = a^2$; $144 - 36 = a^2$

    $108 = a^2$

    $\sqrt{108} = 10.4 = a$

7. $26^2 = 10^2 + b^2$; $b^2 = 676 - 100 = 576$

    $b = \sqrt{576} = 24$

9. The ladder length will be the hypotenuse:

    $25^2 = h^2 + 7^2$

    $h^2 = 625 - 49 = 576$

    $h = \sqrt{576} = 24$ ft

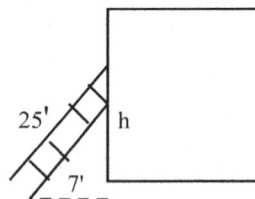

11.

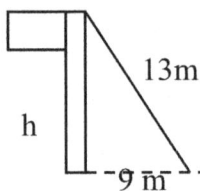

Use the Pythagorean Theorem, letting h = a.

$$h^2 + 9^2 = 13^2$$

$$h^2 = 13^2 - 9^2$$

$$h^2 = 169 - 81 = 88$$

$$h = \sqrt{88} \approx 9.4\,m$$

13.

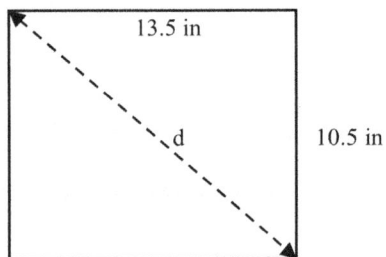

Use the Pythagorean Theorem, letting d = c.

$$a^2 + b^2 = c^2$$

$$13.5^2 + 10.5^2 = c^2$$

$$182.25 + 110.25 = c^2$$

$$292.5 = c^2$$

$$\sqrt{292.5} \approx 17.1 = c$$

This is a 17 in. screen.

15. Lazy Day ←————— 64 mi ————— Jackson

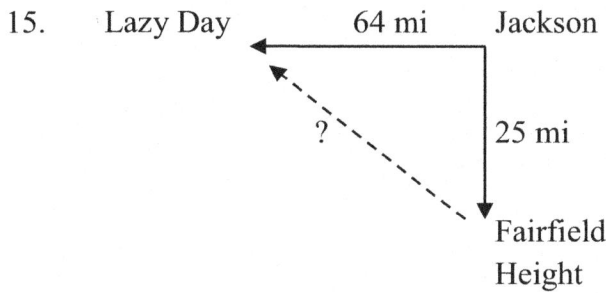

$$a^2 + b^2 = c^2$$
$$64^2 + 25^2 = c^2$$
$$4096 + 625 = c^2$$
$$4721 = c^2$$
$$\sqrt{4721} \approx 68.7 \text{ mi} = c$$

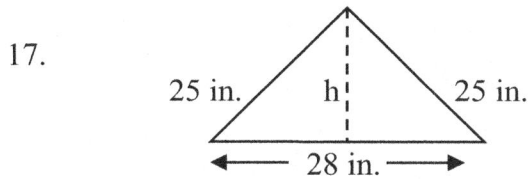

? (25 mi)

Fairfield Height

17.

25 in. ╱ h ╲ 25 in.

←— 28 in. —→

The height (h) divides the triangle into two right triangles with bases of 14 in. Use Pythagorean Theorem to find h.

$$a^2 + b^2 = c^2$$
$$h^2 + 14^2 = 25^2$$
$$h^2 + 196 = 625$$
$$h^2 = 625 - 196 = 429$$
$$h = \sqrt{429} \approx 20.7 \text{ in.}$$

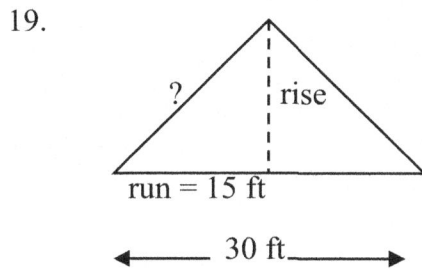

19.

? ╱ rise ╲

run = 15 ft

←——— 30 ft ———→

The slope is given as 11:20. Remember that by definition this is rise over run. Use the slope to find the rise of this roof and then use the Pythagorean Theorem to find the missing side.

$$\frac{11}{20} = \frac{rise}{15}$$
$$(15)(11) = 20x$$
$$165 = 20x$$
$$8.25 = x = rise$$

$$a^2 + b^2 = c^2$$
$$8.25^2 + 15^2 = c^2$$
$$293.0625 = c^2$$
$$\sqrt{293.0625} \approx 17.1 \text{ ft} = c^2$$

21. Use the Pythagorean Theorem to find the base length:

$$a^2 + b^2 = c^2$$
$$10^2 + b^2 = 26^2$$
$$b^2 = 676 - 100 = 576$$

39

$$b = \sqrt{576} = 24 \text{ cm}$$

Perimeter: $P = 10 \text{ cm} + 26 \text{ cm} + 24 \text{ cm} = 60 \text{ cm}$

Area : $A = \frac{1}{2}(24 \text{ cm})(10 \text{ cm}) = 120 \text{ cm}^2$

23. The length of side AC is 8 units, and the length of side BC is 4 units. Use the Pythagorean Theorem to find the length of AB.

$$a^2 + b^2 = c^2$$
$$8^2 + 4^2 = c^2$$
$$64 + 16 = c^2$$
$$80 = c^2$$
$$\sqrt{80} \approx 8.9 = c$$

25. The doorway is a rectangle. In order for the mirror to fit through the doorway, it would be turned diagonally to pass through the doorway. If the diameter of the mirror is longer than the diagonal of the doorway, it will not fit.

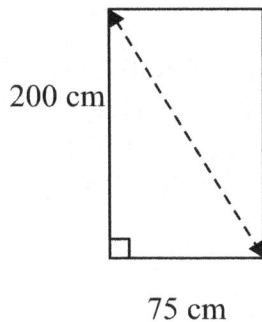

200 cm

75 cm

$$a^2 + b^2 = c^2$$
$$200^2 + 75^2 = c^2$$
$$40{,}000 + 5625 = c^2$$
$$45{,}625 = c^2$$
$$\sqrt{45625} \approx 213.6 = c$$

Since the mirror has a 220cm diameter and the doorway is only 213.6cm on the diagonal, the mirror will not go through the door.

27. The diagonal will be the hypotenuse of an isosceles right triangle with sides *s*.

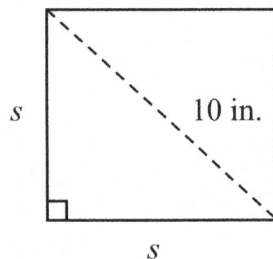

s   10 in.

s

$$a^2 + b^2 = c^2$$
$$s^2 + s^2 = 10^2$$
$$2s^2 = 100$$
$$s^2 = 50$$
$$s = \sqrt{50} \approx 7.1 \text{ in.}$$

40

Practice Set 2-4

1. To make drawings in 2-dimensions that are true representations of what we see in 3-dimensions with our eyes.

3. The term implies that balance or a regular pattern exists

5. horizontal symmetry: B, C, D, E, H, I, O, X.
   vertical symmetry: A, H, I, M, O, T, U, V, W, X, Y.
   F, G, J, K, L, N, P, Q, R, S, and Z do not have horizontal or vertical symmetry

7. H, I, O, S, X, Z.

9. one-third of a full rotation = 120°

11. one-fifth of a full rotation = 72°

13.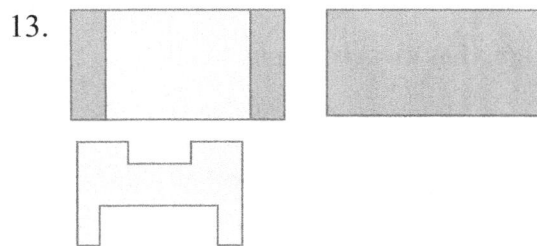

15. $\left(\dfrac{16 \text{ ft}}{1 \text{ in.}}\right)(21 \text{ in}) = 336 \text{ ft}$

17. $\left(\dfrac{3}{4}\right)(175 \text{ cm}) = 131.25 \text{ cm}$

19. $\left(\dfrac{800 \text{ cm}}{1 \text{ cm}}\right)(42.2 \text{ cm}) = 33{,}760 \text{ cm} = 337.6 \text{ m}$ long

    $\left(\dfrac{800 \text{ cm}}{1 \text{ cm}}\right)(8.2 \text{ cm}) = 6560 \text{ cm} = 65.6 \text{ m}$ wide

21. $(8 \text{ cm})\left(\dfrac{5000}{1}\right) = 40{,}00 \text{ cm} = 400 \text{ m}$

23. approximate length of photo = 2.375in.
    $\dfrac{573 \text{ ft}}{2.375 \text{ in}} = 241.263$; or 1 in. = 241 ft
    (answers may vary slightly depending upon accuracy of measurement)

41

25. $(600 \text{ mi})\left(\dfrac{0.5 \text{ in.}}{50 \text{ mi}}\right) = 6 \text{ in.}$

27. $(23 \text{ ft})(1.618) \approx 37.214 \text{ ft} = 37 \text{ ft } 2.5 \text{ in.}$

29. $(45.5\text{cm})(1.618) \approx 73.6\text{cm}; \quad \dfrac{45.5 \text{ cm}}{1.618} \approx 28.1 \text{ cm}$

Practice Set 2-5

1. Pitch refers to how our ears interpret different frequencies of sound. The higher the frequency of the sound wave, the higher pitch we hear.

3. A recording in which the actual wave form is copied as closely as physically (technologically) possible.

5. Frequencies are measured numerically and recorded as numbers in short strings.

7. One octave higher = (261.626 Hz)(2) = 523.252 Hz
   One octave lower = (261.626 Hz)/2 = 130.813 Hz

9. (392 Hz)(1.05946) = 415.3 Hz

11. 4/4 time = 4 quarter notes and each 2 quarter notes = 1 half note, so 2 half notes per measure.

   Or 4/4 time $= \dfrac{4}{4} = \dfrac{2}{4} + \dfrac{2}{4} = \dfrac{1}{2} + \dfrac{1}{2} = 2$ half notes

13. 3/2 time = 3 halves or 3 half notes per measure; $\dfrac{3}{2} = 3\left(\dfrac{1}{2}\right) = 3$ half notes.

15. $\dfrac{3}{4} = \dfrac{1}{4} + \dfrac{1}{4} + x$; x = 1 quarter note missing.

17. $\dfrac{3}{8} = \dfrac{1}{8} + \dfrac{1}{8} + x = \dfrac{2}{8} + x$; x = 1 eighth note missing

19. $\dfrac{2}{4} = \dfrac{1}{2} + x$; x = 0 notes missing

21. y = 5sin220t; (similar to fig. 2-25c)

23. y = 5sin196t; (similar to fig. 2-25c)

25. (group project results will vary)

42

Chapter 2 Review Problems

1. P = 10 + 25 + 6 + 16 + (10 - 6) + (25 - 16) = 70 m
   A = (10)(25 - 16) + (16)(6) = 186 m^2

2. amount of molding = L + H + H + half the circumference of a circle for the curved top
   amt. of molding = 5 + 4.5 + 4.5 + ½(π)(5) ≈ 21.85390 ft
   cost = ($0.55)(21.85390) = 12.01968 = $12.02

3. 180° - 35° - 52° = 93°; Since one angle is > 90°, this is an obtuse triangle.

4. If d = 36 cm, then r = 18 cm.
   A = πr^2 = π(18 cm)^2 ≈ 1017.9 cm^2
   C = πd = π(36 cm) ≈ 113.1 cm

5. A = $\frac{1}{2}$ bh = $\frac{1}{2}$ (18 in.)(10 in.) = 90 in.^{2}
   P = 30 in. + 18 in. + 15 in. = 63 in.

6. Find the total area that includes the circular sidewalk and tree. Then subtract the area of the tree in the center.
   A = π(9.5)^2 - π(1.5)^2 ≈ 276.5 ft^2

7. Use Hero's formula: $s = \frac{6+8+12}{2} = \frac{26}{2} = 13$
   $A = \sqrt{13(13-6)(13-8)(13-12)} = \sqrt{13(7)(5)(1)} = \sqrt{455} \approx 21.3 \, in^2$

8. (a) Here we break the figure into the area of a rectangle plus the area of a triangle by Hero's formula minus the area of each window.
   For Hero's formula: $s = \frac{25+25+42}{2} = 46$
   $A = 42(30) + \sqrt{46(46-25)(46-25)(46-42)} - 3(3.5) - 3(3.5) - (9)(3.5) \approx 1492.357 \approx 1492.4 \, ft^2$

   (b) Since 1 yd^2 = 9 ft^2, divide to convert to square yards.
   $\frac{1492.4 \, ft^2}{9 \, ft^2/yd^2} \approx 165.82 = 166 \, yd^2$
   166 yd^2 ($22.25/ yd^2) = $3693.50

9. The length of the ladder, L, will be the hypotenuse of a right triangle.
   c^2 = a^2 + b^2
   L^2 = 15^2 + 8^2 = 289
   L = $\sqrt{289}$ = 17 feet

10. The first base to third base distance, d, across the middle of the infield will be the hypotenuse of a right triangle.

$$c^2 = a^2 + b^2$$
$$d^2 = 60^2 + 60^2 = 7200$$
$$d = \sqrt{7200} \approx 84.8528 = 84.9 \text{ ft}$$

11. If these side lengths do form a right triangle, the longest side must be the hypotenuse and the lengths of the other sides must correctly fulfill the Pythagorean theorem.

$$c^2 = a^2 + b^2$$
$$12.5^2 = 10^2 + 7.5^2$$
$$156.25 = 100 + 56.25$$
$$156.25 = 156.25 \qquad \text{Therefore, this triangle is a right triangle.}$$

12. Since the base of the isosceles triangle is 24 in., we know that the height from the vertex angle to the base divides the triangle into two right triangles. We use the Pythagorean Theorem to find the missing length of the side of the triangle.

$$c^2 = a^2 + b^2$$
$$s^2 = 12^2 + 18^2 = 468$$
$$s = \sqrt{468} \approx 21.635\ldots = 21.6 \text{ feet}$$

13. a) vertical reflection symmetry
    b) horizontal reflection symmetry
    c) vertical reflection symmetry
    d) neither type of symmetry exists here

14. scaling factor $= \dfrac{50 \text{ cm}}{75 \text{ cm}} = \dfrac{2}{3}$; a scaling factor of 2 to 3

15. actual length $= (4.125 \text{ in.}) \left(\dfrac{32 \text{ in.}}{1 \text{ in.}} \right) = 132$ in. $= 11$ ft

16. the shorter sides $= \dfrac{32.4 \text{ cm}}{1.62} = 20$ cm

17. $(35 \text{ mi}) \left(\dfrac{1 \text{ in.}}{20 \text{ mi}} \right) = 1.75$ in.

18. $\dfrac{1}{4} + \dfrac{1}{8} + x = \dfrac{3}{8} + x$; $\dfrac{5}{8}$ needed so $\dfrac{1}{8} + \dfrac{4}{8}$ (or $\dfrac{1}{2}$) or 1 eighth note and 1 half note

19. Two octaves lower $= (55 \text{ Hz})/2 = 27.5$ Hz

20. $\dfrac{3}{2} = \dfrac{6}{4} = 6$ quarter notes per measure

44

Chapter 2 Test

1. $P = 24 \text{ cm} + 10 \text{ cm} + 26 \text{ cm} = 60 \text{ cm}$

   $A = \frac{1}{2}bh = \frac{1}{2}(10 \text{ cm})(24 \text{ cm}) = 120 \text{ cm}^2$

2. Area of the rectangle – Areas of the two circles = shaded portion

   $lw - 2\pi r^2 = (12 \text{ in.})(6 \text{ in.}) - 2\pi(3 \text{ in.})^2 \approx 72 - 56.5 = 15.5 \text{ in.}^2$

3. Find the circumference of the tablecloth and then multiply by 3.

   $C = \pi d = \pi(6 \text{ ft}) \approx 18.84955 \text{ ft} \cdot 3 \approx 56.5 \text{ ft}$

4. $180° - 25° - 65° = 90°$; right triangle

5. $\quad a^2 + b^2 = c^2$

   $a^2 + 7.2^2 = 7.8^2$

   $a^2 + 51.84 = 60.84$

   $\qquad a^2 = 9$

   $\qquad a = \sqrt{9} = 3 \text{ ft}$

6. Draw a perpendicular height from the peak to the base forming two right triangles. The hypotenuse is 10 cm and the base is 6 cm. Find the height by using the Pythagorean Theorem.

   $a^2 + b^2 = c^2$

   $a^2 + 6^2 = 10^2$

   $a^2 + 36 = 100$

   $\quad a^2 = 64$

   $\quad a = \sqrt{64} = 8 \text{ cm}$

   Now, calculate the area of the triangle with b = 12 cm and h = 8 cm.

   $A = \frac{1}{2}bh = \frac{1}{2}(12 \text{ cm})(8 \text{ cm}) = 48 \text{ cm}^2$

7. Use Hero's formula: $s = \dfrac{2.6 + 3.4 + 4.1}{2} = 5.05$

   $A = \sqrt{5.05(5.05 - 2.6)(5.05 - 3.4)(5.05 - 4.1)} = 4.4038 = 4.4 \text{ ft}^2$

8. (a) $P = 21.5 + (18.5 - 4.25) + (21.5 - 15) + 4.25 + 15 + 18.5 = 80 \text{ ft}$

   $A = 18.5(15) + (18.5 - 4.25)(21.5 - 15) = 370.125 \text{ ft}^2$

   (b) $\dfrac{370.125}{9} = 41.125 \text{ yd}^2$ or 42 yd^2 if you must buy whole square yards

   $\text{Cost} = 41.125(\$24.95/\text{yd}^2) = 1026.06875 \approx \$1026.07$

   or $42 \cdot \$24.95/\text{yd}^2 = \$1047.90$ for whole square yards

45

9. The ladder's length will be the hypotenuse of a right triangle.

$$c^2 = a^2 + b^2$$
$$26^2 = 4.5^2 + b^2$$
$$b^2 = 655.75$$
$$b = \sqrt{655.75} \approx 25.6076 = 25.6 \text{ ft}$$

10. Use the Pythagorean Theorem with the diagonal as the hypotenuse and length as the base.

$$c^2 = a^2 + b^2$$
$$6.5^2 = 5.8^2 + b^2$$
$$b^2 = 8.61$$
$$b = \sqrt{8.61} \approx 2.9342 = 2.9 \text{ cm}$$

11. (a) In one revolution, it will roll a distance equal to the circumference of a circle. The diameter (d) of the tire is 21 in.
$$C = \pi d = \pi(21) \approx 65.973\ldots \approx 65.97 \text{ in.; Convert to feet: } \frac{65.97}{12} \approx 5.498 \text{ ft}$$

    (b) The number of revolutions $= \frac{5280 \text{ ft/mi}}{5.498 \text{ ft}} \approx 960.349$ or a little over 960 revolutions

12. Only letter "O"

13. Letters O and X

14. $(6250 \text{ cm}) \left(\frac{1 \text{ cm}}{2500 \text{ m}} \right) = 2.5 \text{ cm}$

15. $(75 \text{ ft}) \left(\frac{1 \text{ ft}}{87 \text{ ft}} \right) \approx 0.862\ldots\text{ft}$

    Convert feet to inches: $(0.862 \text{ ft}) \left(\frac{12 \text{ in.}}{1 \text{ ft}} \right) = 10.3 \text{ in.}$

16. Since $\frac{64 \text{ in.}}{48 \text{ in.}} = 1.333 \neq 1.62$, the ratio is not equivalent to the Golden Ratio.

17. notebook paper ratio $= \frac{11 \text{ in.}}{8.5 \text{ in.}} = 1.29$

    legal pad ratio $= \frac{14 \text{ in.}}{8.5 \text{ in.}} = 1.65$; The legal pad is closer to the Golden Ratio of 1.62.

18. 2(440Hz) = 880Hz (one octave)
    2(880Hz) = 1760Hz (two octaves)
    2(1760Hz) = 3520 Hz = three octaves above 440Hz

19. $\frac{1}{16} + \frac{1}{16} + \frac{1}{2} + \frac{1}{4} + x = \frac{14}{16} + x = \frac{7}{8} + x$; Therefore, $x = \frac{1}{8}$ or one eighth note is necessary to complete the measure.

20. 3/4 time music has 3 quarter notes per measure and 4/4 time music has 4 quarter notes per quarter. The difference is the number of quarter notes per measure.

Chapter 3: Graphing

Practice Set 3-1

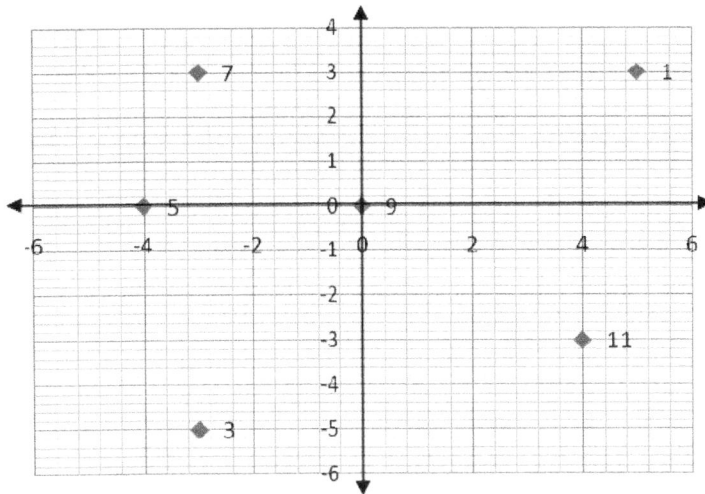

For problems 1-12, all answers are provided and plotted on one graph. Remember, for each (x,y) pair, only one dot appears on the graph. Always start at the origin, (0,0). The 1st number of the pair, the 'x', tells you how many units to move right, R, (if positive) or left, L, (if negative) and, the 2nd number, the 'y', tells you how far to move up, U, (if positive) or down, D, (if negative) from the location in which the first number was placed.

1. (5, 3) = 5 units right from the origin and then 3 units up

3. (-3, -5) = 3 L and 5 D

5. (-4, 0) = 4 L and none up or down (a point on the x-axis line)

7. (-3, 3) = 3 L and 3 U

9. (0, 0) = none R or L and none U or D (the origin)

11. (4, -3) = 4 R and 3 D

For problems 13-20, refer to the diagram on page 56 of the text. Always start at the origin and count units left, L, or right, R, first until you are lined up directly over or under the point, then count up, U, or down, D, until you reach the point. Also see figure 2-1 in the text for a quadrant map. Points on an axis line are not in a particular quadrant.

13. point F is 0 L or R and 0 U or D = (0, 0); i.e., it is at the origin

15. point A is 3 L and 4 U = (-3, 4); in quadrant II

17. point E is 3 L and 3 D = (-3, -3); in quadrant III

19. point G is 0 L or R and 3 D = (0, -3); on the y-axis line (i.e., not in a quadrant)

For numbers 21-29, refer to the diagram on page 56 of the text.

21. Only point C is on the positive x-axis, point F is at the origin where the x-coordinate = 0 and zero is not positive.

23. The y-coordinate is 0, (??,0) points F and C both match this requirement

25. "opposite of" means same absolute value but different signs; only point B has opposite numbers as its coordinates, (1,-1)

27. Is (3, 4) a solution for 2x - 5y = 26?

Substitute: 2(3) - 5(4) = 6 - 20 = -14, not 26. Therefore, (3, 4) is not a solution.

29. Is (3, -2) a solution for 3x + 2y = 5?

Substitute: 3(3) + 2(-2) = 9 + (-4) = 5. Therefore, (3, -2) is a solution.

For problems 30-40, we will find four solutions to complete the chart. First we let $x = 0$ and solve for y, 2nd we let $x = 3$ and solve for y, 3rd we let $y = -2$ and solve for x, and 4th we let $y = 0$ and solve for x.

$x = 0$	$x = 3$	$y = -2$	$y = 0$

31. $x + y = 8$

$0 + y = 8$	$3 + y = 8$	$x + (-2) = 8$	$x + 0 = 8$
$y = 8$	$y = 5$	$x = 10$	$x = 8$
(0, 8)	(3, 5)	(10, -2)	(8, 0)

33. $x - 3y = 12$

$0 - 3y = 12$	$3 - 3y = 12$	$x - 3(-2) = 12$	$x - 3(0) = 12$
$-3y = 12$	$-3y = 9$	$x + 6 = 12$	$x - 0 = 12$
$y = -4$	$y = -3$	$x = 6$	$x = 12$
(0, -4)	(3, -3)	(6, -2)	(12, 0)

35. $x + 4y = 0$

$0 + 4y = 0$	$3 + 4y = 0$	$x + 4(-2) = 0$	$x + 4(0) = 0$
$4y = 0$	$4y = -3$	$x - 8 = 0$	$x + 0 = 0$

50

$y = 0$	$y = -\tfrac{3}{4}$	$x = 8$	$x = 0$
$(0, 0)$	$(3, -\tfrac{3}{4})$	$(8, -2)$	$(0, 0)$

37. $y = x$

$y = 0$	$y = 3$	$-2 = x$	$0 = x$
$(0, 0)$	$(3, 3)$	$(-2, -2)$	$(0, 0)$

39. $-x - 2y = 6$

$-0 - 2y = 6$	$-3 - 2y = 6$	$-x - 2(-2) = 6$	$-x - 2(0) = 6$
$-2y = 6$	$-2y = 9$	$-x + 4 = 6$	$-x - 0 = 6$
$y = -3$	$y = -\tfrac{9}{2} = -4.5$	$-x = 2$	$-x = 6$
		$x = -2$	$-x = 6$
			$x = -6$
$(0, -3)$	$(3, -4.5)$	$(-2, -2)$	$(-6, 0)$

Practice Set 3-2

1. Let the y value = 0 in the equation and solve for x.

3. Calculate the coordinates of another point on the line by assigning either x or y a nonzero value and solving for the matching coordinate.

For problems 5-12 we did three different substitutions into each equation to complete the ordered pairs listed.

5. $x - y = 6$

$x - 0 = 6$	$0 - y = 6$	$x - (-2) = 6$
$x = 6$	$-y = 6$	$x + 2 = 6$
	$y = -6$	$x = 4$
$(6, 0)$	$(0, -6)$	$(4, -2)$

7. $2x + y = 6$

$2x + 0 = 6$	$2(0) + y = 6$	$2x + (-2) = 6$
$2x = 6$	$0 + y = 6$	$2x - 2 = 6$
$x = 3$	$y = 6$	$2x = 8$
		$x = 4$
$(3, 0)$	$(0, 6)$	$(4, -2)$

9. $y = -2$; This means that no matter what x may be chosen to be, y must be -2, so you may just write out the x,y pairs without any calculations: $(3,-2)$, $(0,-2)$, (any real number,-2)

11. $2x + 5y = 10$

$$
\begin{array}{lll}
2x + 5(0) = 10 & 2(0) + 5y = 10 & 2x + 5(-2) = 10 \\
2x + 0 = 10 & 0 + 5y = 10 & 2x - 10 = 10 \\
2x = 10 & 5y = 10 & 2x = 20 \\
x = 5 & y = 2 & x = 10 \\
(5, 0) & (0, 2) & (10, -2)
\end{array}
$$

For problems 13 - 40, follow Example 7 and the rule just above it in the text. In each case below, we chose to find the x-intercept first (let $y = 0$) and the y-intercept second (let $x = 0$).

13. $-x - 2y = 4$

$$
\begin{array}{ll}
-x - 2(0) = 4 & -0 - 2y = 4 \\
-x = 4 & -2y = 4 \\
x = -4 & y = -2 \\
(-4, 0) & (0, -2)
\end{array}
$$

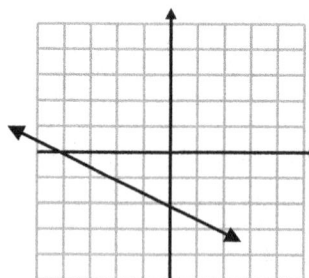

15. $-x - 2y = 5$

$$
\begin{array}{ll}
-x - 2(0) = 5 & -0 - 2y = 5 \\
-x - 0 = 5 & -2y = 5 \\
-x = 5 & y = -\frac{5}{2} \\
x = -5 & \\
(-5, 0) & \left(0, -\frac{5}{2}\right)
\end{array}
$$

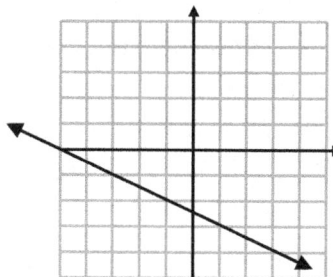

17. $2x + 3y = 6$

$$
\begin{array}{ll}
2x + 3(0) = 6 & 2(0) + 3y = 6 \\
2x = 6 & 3y = 6 \\
x = 3 & y = 2 \\
(3, 0) & (0, 2)
\end{array}
$$

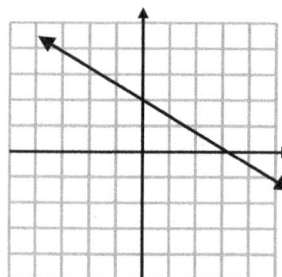

52

19. $3x - 5y = 15$

$$3x - 5(0) = 15 \qquad 3(0) - 5y = 15$$

$$3x = 15 \qquad\qquad -5y = 15$$

$$x = 5 \qquad\qquad y = -3$$

$$(5, 0) \qquad\qquad (0, -3)$$

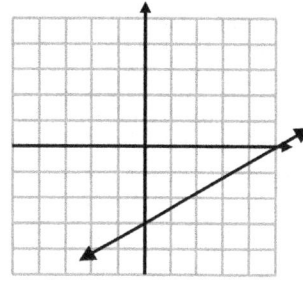

21. $4x - y = 3$

$$4x - 0 = 3 \qquad 4(0) - y = 3$$

$$4x = 3 \qquad\qquad -y = 3$$

$$x = \tfrac{3}{4} \qquad\qquad y = -3$$

$$(\tfrac{3}{4}, 0) \qquad\qquad (0, -3)$$

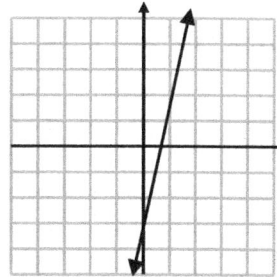

23. $3x = 2y$

$$3x = 2(0) \qquad 3(0) = 2y$$

$$3x = 0 \qquad\qquad 0 = 2y$$

$$x = 0 \qquad\qquad 0 = y$$

$$(0, 0) \qquad\qquad (0, 0)$$

Find another point by letting $y = 3$.

$$3x = 2(3)$$

$$3x = 6$$

$$x = 2 \quad (2, 3) \qquad \text{Draw the line by connecting the origin to (2,3).}$$

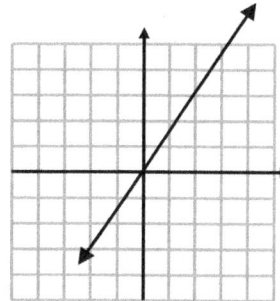

25. $x + y = 0$

$$x + 0 = 0 \qquad 0 + y = 0$$

$$x = 0 \qquad\qquad y = 0$$

$$(0, 0) \qquad\qquad (0, 0)$$

Find another point by letting $y = 1$.

$$x + 1 = 0$$

$$x = -1 \quad (-1, 1)$$

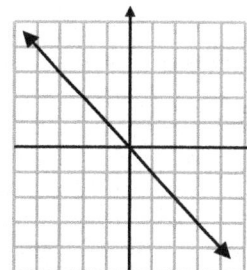

Draw the line by connecting the origin to (-1, 1).

27. $2x + 8 = -4y$; $2x + 4y = -8$

$$2x + 4(0) = -8 \qquad 2(0) + 4y = -8$$
$$2x = -8 \qquad\qquad 4y = -8$$
$$x = -4 \qquad\qquad y = -2$$
$$(-4, 0) \qquad\qquad (0, -2)$$

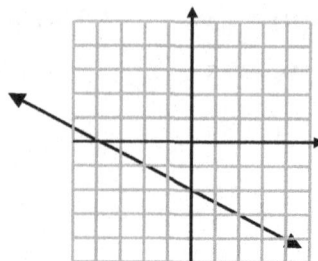

29. $x = 2y + 2$; $x - 2y = 2$

$$x - 2(0) = 2 \qquad 0 - 2y = 2$$
$$x = 2 \qquad\qquad -2y = 2$$
$$\qquad\qquad\qquad y = -1$$
$$(2, 0) \qquad\qquad (0, -1)$$

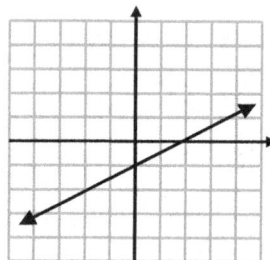

31. $y = 0.2x - 4$; $-0.2x + y = -4$

$$-0.2x + 0 = -4 \qquad -0.2(0) + y = -4$$
$$-0.2x = -4 \qquad\qquad y = -4$$
$$x = 20$$
$$(20, 0) \qquad\qquad (0, -4)$$

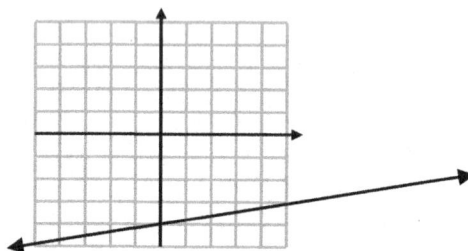

33. $x + \frac{1}{2}y = 3 \qquad\qquad 0 + \frac{1}{2}y = 3$

$$x + \frac{1}{2}(0) = 3 \qquad\qquad \frac{1}{2}y = 3$$
$$x = 3 \qquad\qquad\qquad y = 6$$
$$(3, 0) \qquad\qquad\qquad (0, 6)$$

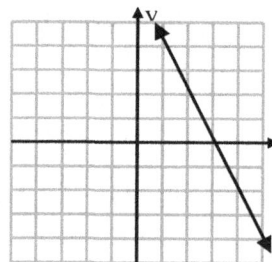

35. $x = -3$

no place for $y = 0$ no y-intercept because

so y can be any   x cannot $= 0$

number, like 0

$(-3, 0)$

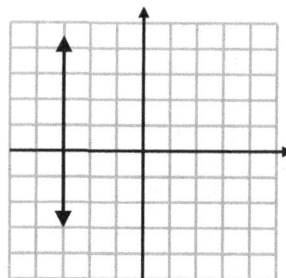

54

37. $3y = 2$; $y = \frac{2}{3}$

     no x-intercept because y cannot = 0

     so x may be any number you choose, like 0 (0, $\frac{2}{3}$)

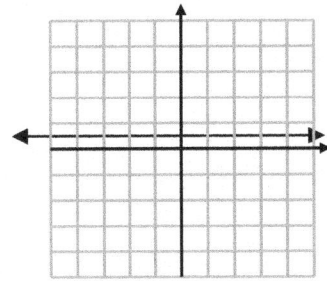

39. $3y + x = 4$

$3(0) + x = 4$	$3y + 0 = 4$
$x = 4$	$3y = 4$
	$y = \frac{4}{3}$
$(4, 0)$	$(0, \frac{4}{3})$

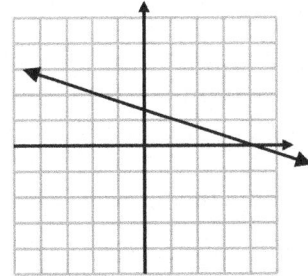

Practice Set 3-3

The slopes of the lines in problems 1-12 were all calculated following Examples 13, 14 and 15.

1. $m = \dfrac{1-(-4)}{6-8} = \dfrac{1+4}{-2} = -\dfrac{5}{2}$

3. $m = \dfrac{0-2}{-3-8} = \dfrac{-2}{-11} = \dfrac{2}{11}$

5. $m = \dfrac{-2-8}{-8-(-4)} = \dfrac{-10}{-4} = \dfrac{5}{2}$

7. $m = \dfrac{3-(-9)}{2.5-1} = \dfrac{12}{1.5} = 8$

9. $\dfrac{7}{1} = \dfrac{8-1}{x-4}$

    $7(x - 4) = 1(7)$ [cross multiply]

    $7x - 28 = 7$

$$7x = 35$$

$$x = 5$$

11. $\dfrac{7}{1} = \dfrac{-6 - y}{3 - 7}$

    $7(-4) = 1(-6 - y)$ [cross multiply]

    $-28 = -6 - y$

    $-22 = -y$

    $22 = y$ [divide both sides by -1]

For numbers 13 - 25, the equations will be arranged in slope-intercept form, y = mx + b, where m is the slope of the line and b is the y-intercept (0,b). These graphs are approximations and you will find much more detailed graphs in the text.

13.   y = -4x + 1

    m = -4, y-intercept = (0, 1)

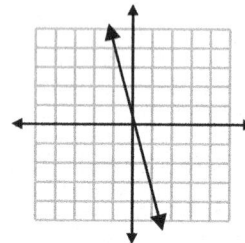

15.   3x + 2y = 6

    2y = -3x + 6

    $y = -\dfrac{3}{2}x + 3$

    $m = -\dfrac{3}{2}$, y-intercept = (0, 3)

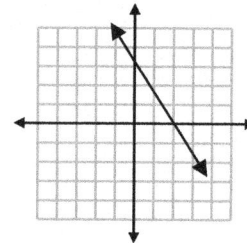

17.   2x - 3y = 4

    -3y = -2x + 4

    $y = \dfrac{2}{3}x - \dfrac{4}{3}$

    $m = \dfrac{2}{3}$, y-intercept = $(0, -\dfrac{4}{3})$ or (0, -1.33)

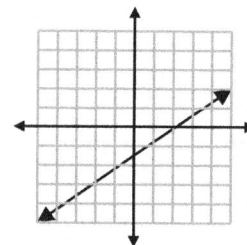

56

19. $4y - x = 10$

$4y = x + 10$

$y = \dfrac{1}{4}x + \dfrac{5}{2}$

$m = \dfrac{1}{4}$, y-intercept = $(0, \dfrac{5}{2}) = (0, 2.5)$

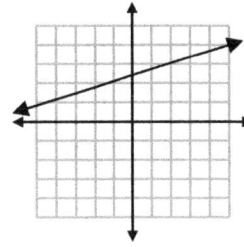

21. $\dfrac{3}{4}x - \dfrac{1}{2}y = \dfrac{5}{8}$

clear fractions, LCD = 8

$8\left(\dfrac{3}{4}x - \dfrac{1}{2}y\right) = 8\left(\dfrac{5}{8}\right)$

$6x - 4y = 5$

$-4y = -6x + 5$

$y = \dfrac{3}{2}x - \dfrac{5}{4}$

$m = \dfrac{3}{2}$, y-intercept = $(0, -\dfrac{5}{4}) = (0, -1.25)$

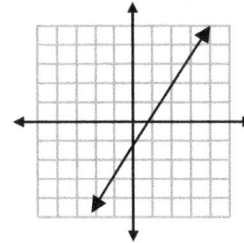

23. $y = -x$

$m = -1$, y-intercept = $(0, 0)$

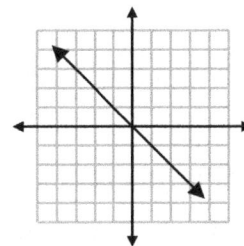

25. $4y = 12 + x$

$y = \tfrac{1}{4}x + 3$

$m = \tfrac{1}{4}$, y-intercept = $(0, 3)$

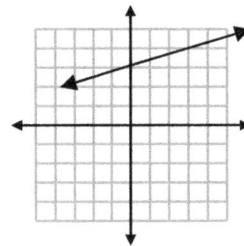

For problems 26-29, first find the slope of the original line. This will be done by placing the equation given into slope-intercept form, $y = mx + b$, and reading the slope, m, directly from the equation. The slope of any line parallel to the original line will have the <u>same slope</u> as the original line's slope.

27. $3x - 8y = 11$

$$-8y = -3x + 11$$

$$y = \frac{3}{8}x - \frac{11}{8} \; ; \; m = \frac{3}{8}$$

29. $-x + 6y = 2$

$$6y = x + 2$$

$$y = \frac{1}{6}x + \frac{1}{3} \; ; \; m = \frac{1}{6}$$

For problems 31-35, first find the slope of the original line. This will be done by placing the equation given into slope-intercept form, $y = mx + b$, and reading the slope, m, directly from the equation. The slope of any line perpendicular to the original line will have a slope that is the <u>negative reciprocal</u> of the original line's slope.

31. $x + y = 3$

$$y = -x + 3 \; ; \; m = -1 \; ; \text{ the slope of a perpendicular line} = 1$$

33. $x = 2$; this is a vertical line and all vertical lines have an undefined slope ; any line perpendicular to a vertical line will be a horizontal line and all horizontal lines have a slope of zero; thus, the slope of a perpendicular line $= 0$

35. $y = 2$; $m = 0$; this is a horizontal line and any line perpendicular to it will be a vertical line; all vertical lines have undefined slopes; thus, the slope of a perpendicular line = undefined

For problems 36-40 we will find the slopes of both lines by placing both lines in slope-intercept form, $y = mx + b$, and then reading the slopes, m, directly from the equations. If the slopes are the same then the lines are parallel. If the slopes are negative reciprocals of each other then the lines are perpendicular. If neither of the above relationships exists, then the lines are neither parallel nor perpendicular.

37. $x + y = 5$ $x - y = -5$

     $y = -x + 5$ $y = x + 5$

     $m = -1$ $m = 1$

These two lines have slopes that are negative reciprocals of one another so they are perpendicular.

39. $y = 2x$ $y = x$

     $m = 2$ $m = 1$

These slopes are not the same nor are they negative reciprocals of each other so these lines are neither parallel nor perpendicular to each other.

Practice Set 3-4

1. Use the slope-intercept form $y = mx + b$ and substitute: $y = -2x + 8$.

3. Use the slope-intercept form and substitute $m = \frac{1}{2}$ and $(x, y) = (6, -2)$ to find b.

$$-2 = \frac{1}{2}(6) + b$$

$$-2 = 3 + b$$

$$-5 = b$$

Now substitute *m* and *b* into the slope intercept form to give $y = \frac{1}{2}x - 5$.

5. Use the slope-intercept form and substitute $m = -2$ and $(x, y) = (-8, -9)$ to find b.

$$-9 = -2(-8) + b$$

$$-9 = 16 + b$$

$$-25 = b$$

Now substitute *m* and *b* into the slope-intercept form to give $y = -2x - 25$.

7. Use the slope-intercept form and substitute $m = \frac{5}{2}$ and $(x, y) = (-1, 4)$ to find b.

$$4 = \frac{5}{2}(-1) + b$$

$$4 = -\frac{5}{2} + b$$

$$4 + \frac{5}{2} = b$$

$$\frac{8}{2} + \frac{5}{2} = b$$

$$\frac{13}{2} = b$$

Now substitute *m* and *b* into the slope intercept form to give $y = \frac{5}{2}x + \frac{13}{2}$.

59

9. Use the slope-intercept form with $m = -\dfrac{1}{3}$ and the x-intercept point being (5, 0).

$$0 = -\frac{1}{3}(5) + b$$

$$0 = -\frac{5}{3} + b$$

$$\frac{5}{3} = b$$

Now substitute m and b into the slope-intercept form to give $y = -\dfrac{1}{3}x + \dfrac{5}{3}$.

For problems 11 - 22, the two points will be used to find the slope using the usual formula. Then the slope and one of the points (we used the first point given in each problem) will be used in the slope-intercept form of the equation of a line.

11. $m = \dfrac{9-3}{7-4} = \dfrac{6}{3} = 2$

Now: $3 = 2(4) + b$

$3 = 8 + b$

$-5 = b$

Substitute the values for m and b: $\quad y = 2x - 5$

13. $m = \dfrac{-2-(-2)}{8-5} = \dfrac{0}{3} = 0$; a horizontal line

Now: $-2 = 0(5) + b$

$-2 = b$

Substitute the values for m and b: $\quad y = 0x - 2$ or $y = -2$

15. $m = \dfrac{2-4}{1-(-1)} = \dfrac{-2}{2} = -1$

Now: $4 = -1(-1) + b$

$4 = 1 + b$

$3 = b$

Substitute the values for m and b: $\quad y = -x + 3$

17. $m = \dfrac{\frac{1}{3} - 1\frac{1}{3}}{5 - 7} = \dfrac{-1}{-2} = \dfrac{1}{2}$

   Now: $1\frac{1}{3} = \frac{1}{2}(7) + b$

   $\qquad 1\frac{1}{3} = \frac{7}{2} + b$

   $\qquad \frac{4}{3} - \frac{7}{2} = \frac{8}{6} - \frac{21}{6} = -\frac{13}{6} = b$

   Substitute the values for *m* and *b*: $y = \dfrac{1}{2}x - \dfrac{13}{6}$

19. $m = \dfrac{-4.5 - 3}{-1 - (-4)} = \dfrac{-7.5}{3} = -2.5$

   Now: $3 = -2.5(-4) + b$

   $\qquad 3 = 10 + b$

   $\qquad -7 = b$

   Substitute the values for *m* and *b*: y = -2.5x – 7

21. $m = \dfrac{1\frac{1}{3} - (-5)}{3 - 3} = \dfrac{6\frac{1}{3}}{0} = undefined$;

   a vertical line where x = constant = 3; in this case, x = 3

23. A line with an undefined slope is a vertical line with x = constant. The given point
   of (1, 4) tells us that x = 1.

Problems 25-36 all relate to either parallel lines (lines having the same slopes) or
perpendicular lines (lines having slopes that are negative reciprocals of one another).
First find the slope by either rearranging the given equation or calculating it. If the lines
are parallel, use the calculated slope. If the lines are perpendicular, find the negative
reciprocal of the given line. Then use the slope-intercept form of the line to write the
equation as you have done in previous problems.

25. 2x + y = 6

   in slope-intercept form: y = -2x + 6 ; m = -2 ; parallel line's slope = -2
   Now: -7 = -2(1) + b

   $\qquad -7 = -2 + b$

   $\qquad -5 = b$

   Substitute the values for *m* and *b*: y = -2x - 5.

61

27. $x - 3y = -4$

   in slope-intercept form: $y = \frac{1}{3}x + \frac{4}{3}$; $m = \frac{1}{3}$; perpendicular line's slope = -3

   Now: $10 = -3(3) + b$

   $\qquad 10 = -9 + b$

   $\qquad 19 = b$

   Substitute the values for *m* and *b*: $y = -3x + 19$

29. $m = \dfrac{11-8}{-1-2} = \dfrac{3}{-3} = -1$; the slope of parallel lines also = -1

   Now: $9 = -1(-2) + b$

   $\qquad 9 = 2 + b$

   $\qquad 7 = b$

   Substitute the values for *m* and *b*: $y = -x + 7$

31. $m = \dfrac{3-1}{1-(-4)} = \dfrac{2}{5}$; the slope of perpendicular lines = $-\dfrac{5}{2} = -2.5$

   Now: $2 = -2.5(-5) + b$

   $\qquad 2 = 12.5 + b$

   $-10.5 = b$

   Substitute the values for *m* and *b*: $y = -2.5x - 10.5$

33. The line x = 4 is a vertical line and thus has an undefined slope. Any line parallel to it will also be a vertical line with x = constant. The given point has x = 3, so this is the constant value of x. The equation is x = 3.

35. The line x = 4 is a vertical line and thus has an undefined slope. Any line perpendicular to it will be a horizontal line with y = constant. The given point has y = 3, so this is the constant value of y. The equation is y = 3.

62

37. The x-intercept = (3, 0) and the y-intercept = (0, -1). From these two points the slope may be found and then use the point-slope form of the equation of a line.

$$m = \frac{0-(-1)}{3-0} = \frac{1}{3}$$

Now: $0 = \frac{1}{3}(3) + b$

$0 = 1 + b$

$-1 = b$

Substitute the values for *m* and *b*: $y = \frac{1}{3}x - 1$

39. Passes through the origin (0, 0) and has m = -1.

Since the line passes through the origin, its y-intercept *(b)* value is 0.

Substitute the values for *m* and *b*: $y = -1x + 0$ or $y = -x$.

Practice Set 3-5

For problems 1 – 12, you will need to locate the y-intercept and determine the slope.

Then substitute the y-intercept (*b*) and the slope (*m*) into the slope-intercept form for an

equation of a line.

1. y-intercept = (0, 0) and the slope $= -\frac{1}{2}$

   Substitute the values for *m* and *b*: $\quad y = -\frac{1}{2}x$

3. y-intercept = (0, 2) and the slope = -2

   Substitute the values for *m* and *b*: $\quad y = -2x + 2$

5. y-intercept = (0, -3) and the slope $= \frac{1}{2}$

   Substitute the values for *m* and *b*: $\quad y = \frac{1}{2}x - 3$

7. y-intercept = (0, 2) and the slope $= \frac{1}{3}$

   Substitute the values for *m* and *b*: $\quad y = \frac{1}{3}x + 2$

9. y-intercept = (0, 0) and the slope = 1

Substitute the values for m and b: $y = x + 0$ or $y = x$

11. y-intercept = 3 and the slope = 0

Substitute the values for m and b: $y = 0x + 3$ or $y = 3$

13   a.   No

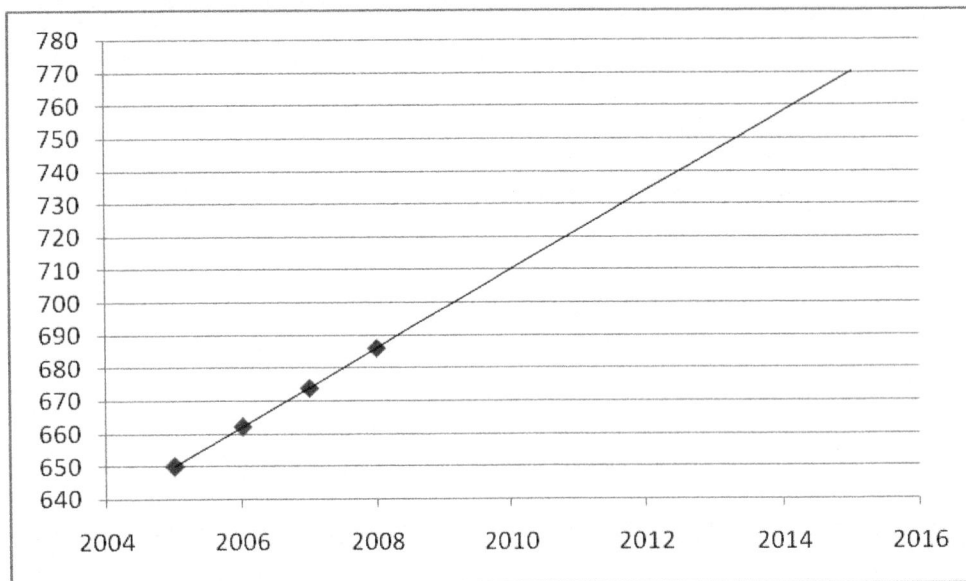

b.   Choose two points to calculate the slope.

(2005, 650) and (2006, 662)

$$m = \frac{662 - 650}{2006 - 2005} = \frac{12}{1} = 12 = \text{ rate of change}$$

So rate of change = 12 students per year

c.   Since 2005 is the initial value, the y-intercept is 650.

Substitute the values of m and b: $y = 12x + 650$

d.   Since the x = number of years after 2005, the x value is 8 for the year 2013.

Substitute x = 8 into the equation $y = 12(8) + 650 = 96 + 650 = 746$

The number of students enrolled in 2013 is predicted to be 746.

15. a.

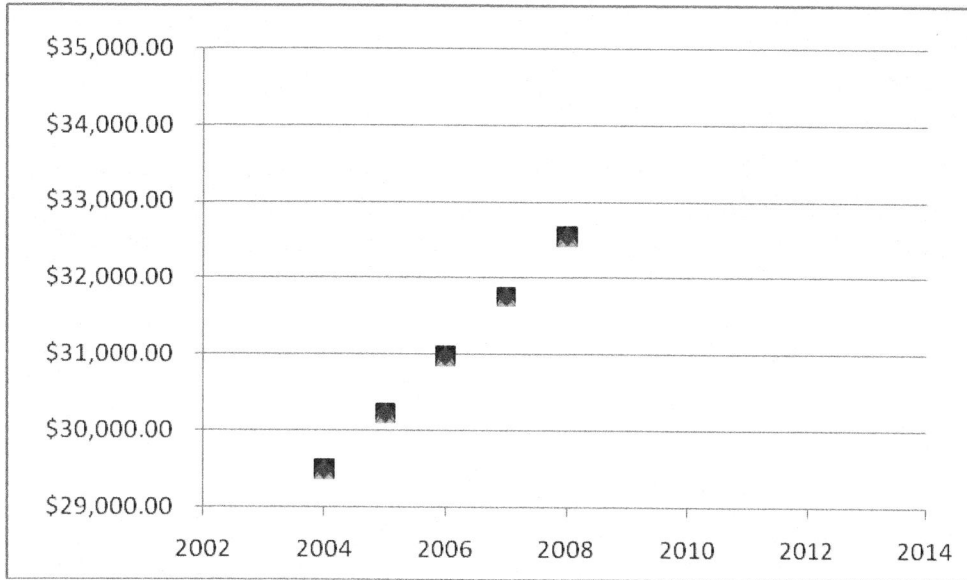

b. The rate of change is not constant so you cannot calculate a constant slope.
c. Since the rate of change is not constant, you cannot create a linear equation.
d. From the graph, it would appear that the salary will be more than $35,000.
Since the graph is not linear, this is a general estimate.

17. Choose two points to find the slope or rate of change such as (2006, 119145) and

(2007, 111892).

$$m = \frac{111892 - 119145}{2007 - 2006} = \frac{-7253}{1} = -7253 = \text{rate of change}$$

This means a reduction of 7253 tons per year is the rate of change.

No, because a constant rate for the next 7 years would be improbable since there

is only so much refuse that can be recycled thus reducing the total amount.

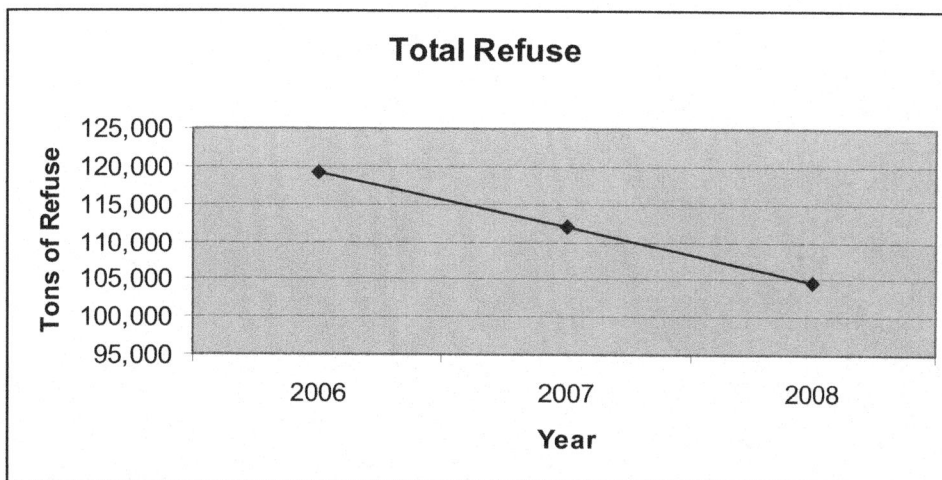

65

19. These rates do not have a linear trend since the changes in price are not consistent throughout the years.

Chapter 3 Review Problems

For problems 1 - 4, all points are to be plotted on one graph. Remember, for each (x,y) pair, only one dot appears on the graph. Always start at the origin, (0,0). The 1st number of the pair, the 'x', tells you how many units to move right, R, (if positive) or left, L, (if negative) and, the 2nd number, the 'y', tells you how far to move up, U, (if positive) or down, D, (if negative) from the location in which the first number was placed. See figure 2-1 in the text for a quadrant map. Points on an axis line are not in any quadrant.

1. $(2,4) = 2$ R and 4 U ; in quadrant I

2. $(-3,4) = 3$ L and 4 U ; in quadrant II

3. $(-2,-1) = 2$ L and 1 D ; in quadrant III

4. $(0,6) =$ none R or L and 6 U ; on the y-axis

5. If you substitute 1 for x and -4 for y into the equation the result will be 6 if this pair of numbers is a solution: $2x - y = 6$

$$2(1) - (-4) = 2 + 4 = 6 \text{, so } (1, -4) \text{ is a solution.}$$

In problems 6-10, substitute the value given in the (x, y) pair and solve for the other value.

6. $2x - y = 8$; $(-2,?)$
$2(-2) - y = 8$
$-4 - y = 8$
$-y = 12$
$y = -12$; $(-2, -12)$

7. $y = -7x + 2$; $(6,?)$
$y = -7(6) + 2$
$y = -42 + 2$
$y = -40$; $(6, -40)$

8. $3x + 2y = 8$; $(?,4)$
$3x + 2(4) = 8$
$3x + 8 = 8$
$3x = 0$
$x = 0$; $(0, 4)$

66

9. $y = 7$; $(?, 7)$

   This is a horizontal line and the values of x can be any real number.

   (any real number, 7)

10. $5x - 3y = 11$; $(0, ?)$

    $5(0) - 3y = 11$

    $-3y = 11$

    $y = -11/3$; $(0, -11/3)$

For problems 11-16, follow Example 7 and the rule just above it in the text. In each case below, we chose to find the x-intercept first (let y = 0) and the y-intercept second (let x = 0).

11. $y = \frac{1}{3}x - 1$; $\frac{1}{3}x - y = 1$

$\frac{1}{3}x - 0 = 1$	$\frac{1}{3}(0) - y = 1$
$\frac{1}{3}x = 1$	$-y = 1$
$x = 3$	$y = -1$
$(3, 0)$	$(0, -1)$
$m = \frac{1}{3}$	

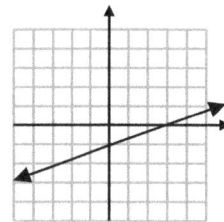

12. $x = -4$

no place for y = 0	this line is a vertical
so y may be any number	line and thus has
you choose, like 0	no y-intercept
$(-4, 0)$	
$m = $ undefined	

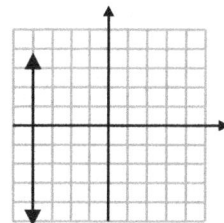

13. $2x + y = 5$

$2x + 0 = 5$	$2(0) + y = 5$
$2x = 5$	$y = 5$
$x = 2.5$	
$(2.5, 0)$	$(0, 5)$
$m = -2$	

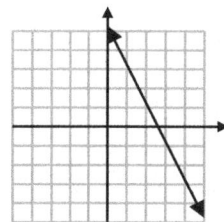

67

14. $-3x + 2y = 6$

$$-3x + 2(0) = 6 \qquad -3(0) + 2y = 6$$
$$-3x = 6 \qquad 2y = 6$$
$$x = -2 \qquad y = 3$$
$$(-2, 0) \qquad (0, 3)$$
$$m = \tfrac{3}{2}$$

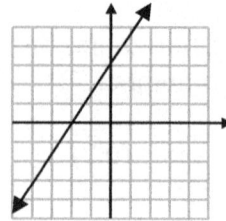

15. $x - 3y = -3$

$$x - 3(0) = -3 \qquad 0 - 3y = -3$$
$$x = -3 \qquad -3y = -3$$
$$\qquad \qquad y = 1$$
$$(-3, 0) \qquad (0, 1)$$
$$m = \tfrac{1}{3}$$

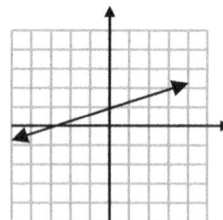

16. $y = 3$ is a horizontal line

no x-intercept because y cannot = 0

and x may be any number you choose, such as 0

which gives (0,3) as a point on the line (y-intercept)

$m = 0$

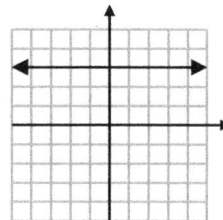

The slopes of the lines in problems 17-20 were all calculated following Examples 13, 14 and 15.

17. $m = \dfrac{4-7}{1-2} = \dfrac{-3}{-1} = 3$

18. $m = \dfrac{4-(-10)}{-2-5} = \dfrac{14}{-7} = -2$

19. $m = \dfrac{6-8}{-3-(-3)} = \dfrac{-2}{0} = \text{undefined}$

20. $m = \dfrac{6-(-1)}{-4-(-6)} = \dfrac{6+1}{-4+6} = \dfrac{7}{2}$

In problems 21-25, the equation will first be written using the slope-intercept form.

21. Use the slope-intercept form: $y = mx + b$

$$y = 2x + (-3) \text{ or } 2x - y = 3$$

22. First find the slope of the line then use the point-slope form:

$$m = \dfrac{6-(-4)}{-2-(-3)} = \dfrac{6+4}{-2+3} = \dfrac{10}{1} = 10$$

Next use the slope-intercept form to find b.

$6 = 10(-2) + b$

$6 = -20 + b$

$26 = b$

Now substitute m and b into the slope-intercept form to give:

$$y = 10x + 26 \text{ or } 10x - y = -26 \text{ (in general form)}$$

23. Use the slope-intercept form with the given slope and point to find b.

$4 = -2(3) + b$
$4 = -6 + b$
$10 = b$

Now substitute m and b into the slope-intercept form to give:

$$y = -2x + 10 \text{ or } 2x + y = 10 \text{ (in general form)}$$

24. Find the slope of the line given and the negative reciprocal of that line's slope will be the slope of the line we are looking for. Then use the slope-intercept form:
$y = 2x - 5$ is written in slope-intercept form, so the slope is 2 and the negative reciprocal of 2 is -½.

$$5 = -\tfrac{1}{2}(-2) + b$$

$$5 = 1 + b$$

$$4 = b$$

Now substitute m and b into the slope-intercept form to give: $y = -\tfrac{1}{2}x + 4$.
Now multiply by 2 to clear the fraction if you want to write the equation in general form $x + 2y = 8$.

25. Parallel lines have the same slope so we first find the slope of the given line and then use the slope-intercept form:

$$3x - 4y = 9$$

$$-4y = -3x + 9$$

$$y = \tfrac{3}{4}x - \tfrac{9}{4} \text{ ; so the slope of the line is } \tfrac{3}{4}$$

Now substitute m and b into the slope intercept form to give: $y = \tfrac{3}{4}x + 4$

If you want to write the equation in general form, then rewrite as $\tfrac{3}{4}x - y = -4$. Now multiply by 4 to clear the fraction to give $3x - 4y = -16$.

26. The y-intercept is (0, -1) and the slope is $\tfrac{1}{2}$.

Substitute m and b into the slope-intercept form: $y = \tfrac{1}{2}x - 1$

If you want to write the equation in general form then rewrite as $-\tfrac{1}{2}x + y = -1$. Now multiply by -2 to clear the fraction to give $x - 2y = 2$.

27. The y-intercept is (0, 4) and the slope is $-\tfrac{1}{2}$.

Substitute m and b into the slope-intercept form: $y = -\tfrac{1}{2}x + 4$

If you want to write the equation in general form, then rewrite as $\tfrac{1}{2}x + y = 4$. Now multiply by 2 to clear the fraction to give $x + 2y = 8$.

Chapter 3 Test

For problems 1-3, all points are to be plotted on one graph. Remember, for each (x, y) pair, only one dot appears on the graph. Always start at the origin, (0, 0). The 1st number of the pair, the 'x', tells you how many units to move right, R, (if positive) or left, L, (if negative) and, the 2nd number, the 'y', tells you how far to move up, U, (if positive) or down, D, (if negative) from the location in which the first number was placed. See figure 2-1 in the text for a quadrant map. Points on an axis line are not in any quadrant. See back of textbook for the graph.

1. (-2, 3) = 2 L and 3 U ; in quadrant II

2. (3, -1) = 3 R and 1 D ; in quadrant IV

3. (-3, -4) = 3 L and 4 D ; in quadrant III

4. If you substitute -3 for x and 5 for y into $3x + 2y$ is the result -1?

    $3(-3) + 2(5) = -9 + 10 = 1$; Since it is not -1, (-3, 5) is not a solution.

For problems 5-9 there are several ways to proceed. We placed the equations in slope-intercept form to find the slope and y-intercept, then solved for the x-intercept. To find the x-intercept let $y = 0$ and the y-intercept is read directly from the equation. The graphs use the intercepts.

70

5. $-4x + y = -4$
    $\qquad y = 4x - 4$
    $m = 4$
    y-intercept = (0, -4)

    solving for the x-intercept: $0 = 4x - 4$
    $\qquad\qquad\qquad\qquad\qquad -4x = -4$
    $\qquad\qquad\qquad\qquad\qquad\quad x = 1 \; ; (1, 0)$

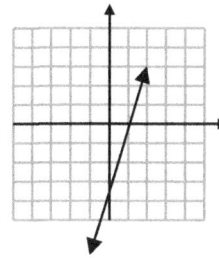

6. $2x + y = -4$
    $\qquad y = -2x - 4$
    $m = -2$
    y-intercept = (0, -4)
    solving for the x-intercept: $0 = -2x - 4$
    $\qquad\qquad\qquad\qquad\qquad 2x = -4$
    $\qquad\qquad\qquad\qquad\qquad\; x = -2 \; ; (-2, 0)$

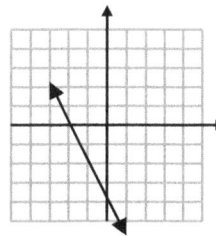

7. $x = 5$

    this is a vertical line
    m = undefined
    y-intercept = none
    x-intercept = (5, 0)

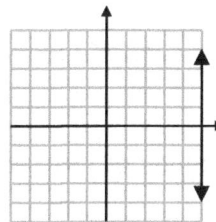

8. $x + y = 3$

    $\qquad y = -x + 3$
    $m = -1$
    y-intercept = (0, 3)
    solving for the x-intercept: $0 = -x + 3$
    $\qquad\qquad\qquad\qquad\qquad\; x = 3 \; ; (3, 0)$

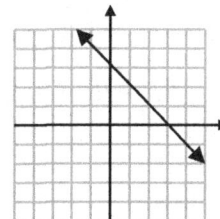

71

9. y = -2

    this is a horizontal line
    m = 0
    y-intercept = (0, -2)
    x-intercept = none

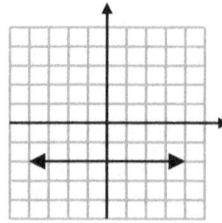

10. $m = \dfrac{y_2 - y_1}{x_2 - x_1} = \dfrac{-2 - (-1)}{-3 - 2} = \dfrac{-1}{-5} = \dfrac{1}{5}$

11. A line perpendicular to a line with slope -3 would have slope $\dfrac{1}{3}$ since perpendicular

    lines have slopes that are negative reciprocals of each other.

12. All vertical lines have *undefined* slope.

13. In the line y = x, *m* (the coefficient of x) has a value of 1.

14. Use the slope-intercept form to find the value of b using the given information.

    6 = -3(-1) + b
    6 = 3 + b
    3 = b

    Now substitute *m* and *b* into the slope-intercept form to give: y = -3x + 3 or in general

    form: 3x + y = 3

15. Use the slope-intercept form to find the value of b using the given information.

    1 = ½ (-6) + b
    1 = -3 + b
    4 = b

    Now substitute *m* and *b* into the slope-intercept form to give: y = ½x + 4 or in general

    form: ½x - y = -4. Multiplying by the LCD gives x - 2y = -8.

16. First find the slope.

    $m = \dfrac{-4 - 2}{0 - 1} = \dfrac{-6}{-1} = 6$

    Now use the slope-intercept form to find the value of b.

    -4 = 6(0) + b

    -4 = b

    Now substitute *m* and *b* into the slope-intercept form to give: y = 6x - 4 or in general

    form: 6x - y = 4

72

17. First find the given line's slope. The slope of the perpendicular line we want will be the negative reciprocal of that slope. Then, use the slope-intercept form.

$$2x - y = 3$$

$$y = 2x - 3 \qquad m = 2 \text{ so the slope of our line will be } -\tfrac{1}{2}$$

Now substitute m and b into the slope-intercept form to give: $y = -\tfrac{1}{2}x + 4$

To write in general form, reorder as $\tfrac{1}{2}x + y = 4$ and multiply by 2 to clear the fraction which gives $x + 2y = 8$.

18. Parallel lines have the same slope so our line will have the same slope as the given line. Find the slope of the given line. $x = 5$ is a vertical line so its slope is undefined; thus, our line must also be vertical and can only have one value for x. In the given point, (2, -3), x is 2, so our line is $x = 2$.

19. The y-intercept = (0, 0) and the slope = $-\tfrac{3}{4}$.

Substitute m and b into the slope-intercept form of the equation. $\qquad y = -\tfrac{3}{4}x$

To write in general form, rewrite as $\tfrac{3}{4}x + y = 0$ and then multiply by 4 to clear the fraction to give $3x + 4y = 0$.

20. The y-intercept = (0, 0) and the slope = $\tfrac{3}{4}$.

Substitute m and b into the slope-intercept form of the equation. $\qquad y = \tfrac{3}{4}x$

To write in general form, rewrite as $-\tfrac{3}{4}x + y = 0$ and then multiply by -4 to clear the fraction to give $3x - 4y = 0$.

73

Chapter 4: Functions

Practice Set 4-1

1. A function is a relation or rule in which, for each input value, there is exactly one output value. The independent variable is the input and is the variable that we control. The dependent variable is the output and its value is a result of the original choice of the value of the independent variable.

3. a) On April 23, the table gives a high temperature of 68°.

    b) The daily high temperature is a function of the date chosen. In other words, the high temperature depends on which date you choose. However, the date does not depend on the high temperature chosen because several dates could have the same high temperature.

    c) The dependent variable is the high temperature.

    d) The domain is the set of dates {20, 21, 22, 23, 24, 25, 26}.

5. a) The independent variable is the year (x-axis) and the dependent variable is the population (y-axis).

    b) The domain (values of x) are 1975-2005. The range (values of y) are 4.2 billion - 6.4 billion.

7. The grade, G, is a function of the identification number, N, since there is only one grade per student.

9. As the weight of a bag increases , the price of the bag increases. Therefore, the price is dependent on the weight of the potatoes.

11. As the slope of a hill increases, the speed of the scooter going uphill decreases. Therefore, the speed is dependent on the steepness of the hill.

13. As a person's age increases, his target heart rate when exercising decreases so the target heart rate when exercising is dependent on a person's age.

15. 9:45 pm

17. January 2 and December 30

19. 10:15 pm is the latest and 2:50 pm is the earliest

21.

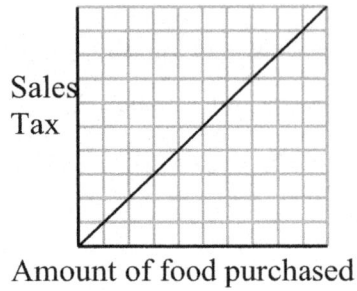

Sales Tax

Amount of food purchased

23. % of homes with a computer

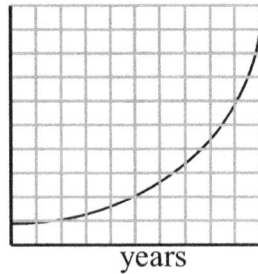

years

25.

height

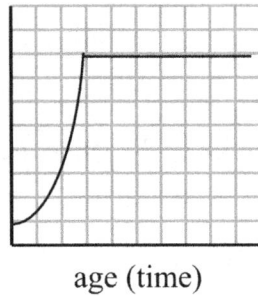

age (time)

27. d = 315 feet at a speed of 70 mph;

The graph is not a straight line. The faster the speed, the longer the stopping distance.

Stopping Distance (in ft)

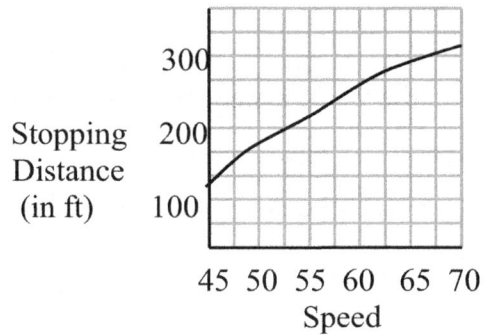

300

200

100

45 50 55 60 65 70
Speed

29.

Interest Rate	5.5%	5.75%	6%	6.25%	6.5%	6.75%	7%
Payment	$709.74	$729.47	$749.44	$769.65	$790.09	$810.75	$831.63

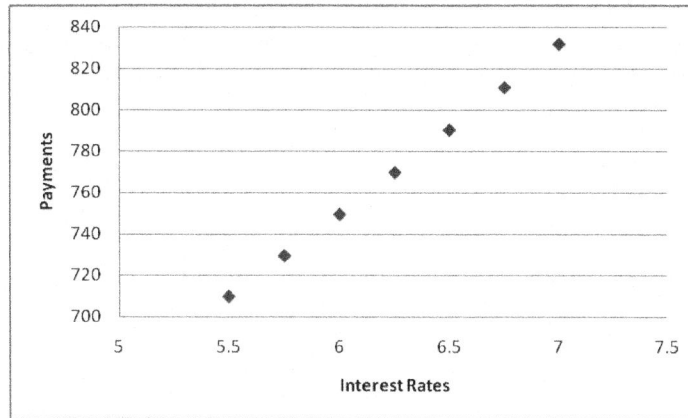

The graph looks similar to a straight line but is not perfectly linear.

Practice Set 4-2

1. $g(0) = 3(0) - 5 = -5$

3. $g(10) = 3(10) - 5 = 30 - 5 = 25$

5. $y = 2x + 1$

   $f(x) = 2x + 1$; Replace y with $f(x)$

   $f(-2) = 2(-2) + 1 = -3$

7. $3x^2 - y = 1$; Solve for y.

   $-y = -3x^2 + 1$ Divide by -1 and then replace y with $f(x)$.

   $f(x) = 3x^2 - 1$

   $f(-2) = 3(-2)^2 - 1 = 3(4) - 1 = 12 - 1 = 11$

9. $5x - 2y = 8$; Solve for y.

   $-2y = -5x + 8$ Divide by -2 and replace y with $f(x)$.

   $f(x) = \dfrac{5}{2}x - 4$

   $f(-2) = \dfrac{5}{2}(-2) - 4 = -5 - 4 = -9$

11. $f(2) = -16(2)^2 - 80(2) + 800 = -16(4) - 80(2) + 800 = -64 - 160 + 800 = 576$

This number tells us that 2 sec after the ball is thrown, it is 576 ft above the ground.

13. $h(46.5) = 2.32(46.5) + 65.53 = 107.88 + 65.53 = 173.41$ cm

15. Since the equation uses cm, we must first convert his height from m to cm.

$$1.83 \text{ m} = 183 \text{ cm}$$

Now we use this number in the function to solve for x (length of the femur).

$$h(x) = 2.32x + 65.53$$

$$183 \text{ cm} = 2.32x + 65.53$$

$$183 - 65.53 = 2.32x + 65.53 - 65.53$$

$$117.47 = 2.32x$$

$$50.63 \text{ cm} = x$$

17. a) $A(n) = 2^n$

$A(30) = 2^{30} = 1,073,741,824$ pennies

b) $1,073,741,824$¢ $= \$10,737,418.24$

19. a) (128 tickets)($7.50 per ticket) = $960

b) $M(n) = \$7.50n$

c) domain = 0 - 200 people; (any number of people from no one to full capacity may attend the movie)

range = $0 - $1500; (amount of money earned from no one attending to full capacity)

21. Recommended heartbeats for a 20-year old would be f(20).

$f(20) = -0.85(20) + 187 = 170$

$f(60) = -0.85(60) + 187 = 136$;   These two values of the function suggest that as a person gets older, his target heart rate decreases.

23. Reading the table for a value of 3 in the left column, f(3) = $29,080.

This means that a person with 3 teaching years experience will make an annual salary of $29,080. The independent variable is the number of years of experience and the dependent variable is the salary that person earns based on experience.

25. $f(4) = 0721$ which means that the sun rises at 7:21am on January 4th; $g(10) = 1654$ which means that the sun sets on January 10th at 1654 hours or 4:54pm; the

78

independent variable is the date and the dependent variables are the sunrise and sunset times.

27. a) Locate the number 5 on the horizontal axis and then the y-value or the point at that location: $f(5) = 75$ mi.

    b) Using the graph, the value is approximately 130 miles (127.5 using the formula)

29. Locate the value 100 on the horizontal axis and find the y-value associated with that location: $f(100) = \$30$.

    This means that it cost $30 to rent this car and drive 100 miles. The domain is numbers ≥ 0 (distances can't be negative) and the range is numbers $\geq \$25$ (minimum amount just to rent the car).

Practice Set 4-3

1. a)

x	$f(x) = 1.75x$	$f(x)$
1	$1.75(1) =$	$1.75
3	$1.75(3) =$	$5.25
5	$1.75(5) =$	$8.75
7	$1.75(7) =$	$12.25
9	$1.75(9) =$	$15.75

   b) initial value is 0; rate of change is $1.75 per mile

   c) $f(x) = \$1.75x$ where x = miles

   d) $f(8) = 1.75(8) = 14$ which means that an 8 mile taxi ride will cost $14.00

3. a) Based on the linear function model $f(x)$ = initial value + (rate of change)(x) the initial value of this function is 5.5 million.

   b) The rate of change is $0.02 = 2\%$ per year

   c) $P(5) = 5.5 + 0.02(5) = 5.6$ million

5. a) $C(x) = \$525 + \$17.50x$ where x = number of guests

   b) $C(200) = 525 + 17.50(200) = \$4025$

   c) We need to use the function to solve for x, the number of guests if the cost is $3150.

$$C(x) = 525 + 17.50(x)$$

$$3150 = 525 + 17.50x$$

$$3150 - 525 = 525 - 525 + 17.50x$$

79

$$2625 = 17.50x$$

$$150 \text{ guests} = x$$

7. a) $\dfrac{1800-100}{5}$ = \$340/year loss of value or rate of change of -\$340 per year

   b) V(x) = initial value + (rate of change)(x) = 1800 - 340x where x = years

   c) V(4) = 1800 - 340(4) = 440 which means that after 4 years, the computer is worth only \$440

9. First use two ordered pairs to find the slope, or rate of change.

$$m = \frac{y_2 - y_1}{x_2 - x_1} = \frac{-2-(-1)}{10-5} = \frac{-1}{5} = -\frac{1}{5}$$

Now use the slope-intercept form to find b and write the equation. Use the point (5, -1).

$$y = mx + b$$

$$-1 = -\frac{1}{5}(5) + b$$

$$-1 = -1 + b$$

$$0 = b$$

$$y = -\frac{1}{5}x + 0 \text{ or } f(x) = -\frac{1}{5}x$$

11. First use two ordered pairs to find the slope, or rate of change

$$m = \frac{5-5}{-1-2} = \frac{0}{-3} = 0$$

Now use the slope-intercept form to find b and write the equation. Use the point (2, 5).

$$y = mx + b$$

$$5 = 0(2) + b$$

$$5 = b$$

$$y = 0x + 5 \text{ or } f(x) = 5$$

13. a) rate of change $= \dfrac{650-450}{2006-1996} = \dfrac{200}{10} = +20/\text{year}$

   b) f(x) = 450 + 20x where x = years after 1996

   c) f(15) = 450 + 20(15) = 750 which means that in the year 2011, the predicted population is 750 people.

15. a) rate of change $= \dfrac{1,000,000 - 600,000}{2008 - 2003} = \dfrac{400,000}{5} = \$80,000/\text{year}$

b) f(x) = \$600,000 + \$80,000x   where x = number of years after 2003

c) Use the function to solve for x, allowing the \$1,500,000 figure to be f(x).

$$f(x) = \$600,000 + \$80,000x$$

$$\$1,500,000 = \$600,000 + \$80,000x$$

$$\$1,500,000 - \$600,000 = \$600,000 - \$600,000 + \$80,000x$$

$$\$900,000 = \$80,000x$$

$$11.25 = x$$

That means the company's sales are predicted to reach \$1,500,000 in 11.25 years or during the year 2015.

17. a) rate of change $= \dfrac{4115 - 3536}{2006 - 2005} = \dfrac{579}{1} = \$579/\text{year}$

b) f(x) = \$3536 + \$579x   where x = number of years after 2005

c)

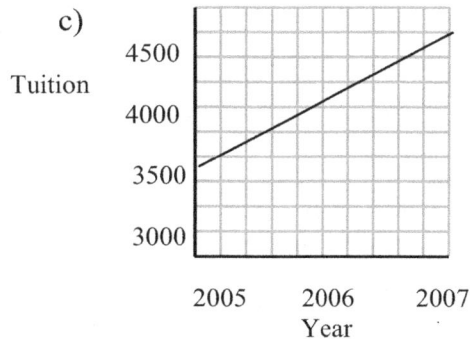

d) f(6) = 3536 + 579(6) = \$7010

19. a) f(x) = 100 + 5x where x = inches over 60 inches   or

f(x) = 100 + 5(x - 60) where x = woman's height in inches

b) f(7) = 100 + 5(7) = 135 lbs

f(67) = 100 + 5(67 - 60) = 135 lbs

Practice Set 4-4

1. direct - As the diameter increases, there is a corresponding increase in circumference.

to check out.

3. inverse - As the amount of water coming in from the hose increases, the time it takes
    to fill the swimming pool decreases.

5. direct - As the amount of water used in a month increases, the water bill increases.

7. $y = kz$

9. $a = kbc$

11. $d = kef^3$

13. $m = \dfrac{k\sqrt{n}}{p^3}$

15. $y = kx$

$8 = k(24)$

$k = \dfrac{24}{8} = \dfrac{1}{3}$

So, $y = \left(\dfrac{1}{3}\right)x$ is the working formula

$y = \left(\dfrac{1}{3}\right)(36) = 12$

17. $y = \dfrac{k}{x}$

$9 = \dfrac{k}{6}$

$k = 54$

So, $y = \dfrac{54}{x}$ is the working formula

$y = \dfrac{54}{36} = 1.5$

19. $y = k\sqrt{x}$

$24 = k\sqrt{16}$

$24 = 4k$

$k = 6$

So, $y = 6\sqrt{x}$ is the working formula.

$y = 6\sqrt{36} = 6 \cdot 6 = 36$

82

21. $y = kxz$

    $150 = k(3)(5^2)$

    $150 = 75k$

    $k = 2$

So, $y = 2xz^2$ is the working formula.

    $y = 2(12)(8^2) = 1536$

23. $p = \dfrac{kq}{r^2}$

    $40 = \dfrac{k(20)}{4^2}$

    $k = 32$

So, $p = \dfrac{32q}{r^2}$ is the working formula.

    $p = \dfrac{32(24)}{8^2} = 12$

25. $a = kr^2$

    $78.5 = k(5)^2$

    $78.5 = 25k$

    $k = 3.14$

So, $a = 3.14r^2$ is the working formula.

    $a = 3.14(7)^2 = 3.14(49) = 153.86 \text{ cm}^2$

27. $d = kt^2$

    $27 = k(3)^2$

    $27 = 9k$

    $k = 3$

So, $d = (3)(t^2)$ is the working formula.

    $d = (3)(2)^2 = 12 \text{ ft}$

29. $p = k\sqrt{l}$

$\quad 0.5 = k\sqrt{16}$

$\quad\quad k = 0.125$

So, $p = 0.125\sqrt{l}$ is the working formula.

$\quad 0.75 = 0.125\sqrt{l}$ [divide both sides by 0.125]

$\quad \dfrac{0.75}{0.125} = \sqrt{l}$

$\quad\quad 6 = \sqrt{l}$ [square both sides to eliminate the square root]

$\quad 36\,in = l$

31. $f = \dfrac{k}{d^2}$

$\quad 20 = \dfrac{k}{(2)^2}$

$\quad\quad k = 80$

So, $f = \dfrac{80}{d^2}$ is the working formula.

$\quad 5 = \dfrac{80}{d^2}$

$\quad 5d^2 = 80$

$\quad d^2 = \dfrac{80}{5}$

$\quad d^2 = 16$

$\quad\quad d = 4\,in$ [take the square root of both sides to eliminate the square]

33. $w = \dfrac{k}{d^2}$

$\quad 180 = \dfrac{k}{(4000)^2}$

$\quad\quad k = 2{,}880{,}000{,}000$

So, $w = \dfrac{2{,}880{,}000{,}000}{d^2}$ is the working formula.

$$w = \frac{2880000000}{(4500)^2}$$

$$w = 142.2 \text{ lb}$$

35. $P = \dfrac{kV^2}{R}$

$$180w = \frac{k(90)^2}{45}$$

$$(180)(45) = k(90)^2 \qquad \text{[cross multiply]}$$

$$\frac{180(45)}{8100} = 1 = k \qquad \text{[divide both sides by 8100]}$$

So, $P = \dfrac{(1)(V^2)}{R}$

$$P = \frac{(1)(120)^2}{30} = 480 \; W$$

37. $e = ktl$

$$2.7 = k(225)(100)$$

$$2.7 = k(22500)$$

$$k = 0.00012$$

So, $e = (0.00012)tl$ is the working formula.

$$e = (0.00012)(150)(40) = 0.72 \text{ in}$$

39. a) direct variation

b) $\dfrac{4}{3}\pi$

c) $V(9) = \dfrac{4}{2}\pi(9)^3 = 972\pi^3 = 3053.6 \; in^3$

Practice Set 4-5

Note: for problems 1 through 5, the graphs are shown as drawn using the graphing function on a graphing calculator and the roots (x-axis intercept points), if any, and maximum or minimum points are found using the CALC function of a graphing

85

calculator as explained on page 129 in the text. The roots may also be found by factoring or using the quadratic formula. Your graphs should match those in the text.

1.   minimum = -4
     roots = -1 and –5

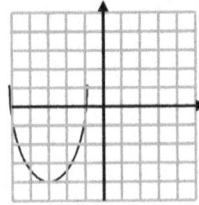

3.   minimum = -2.3
     roots = -1 and –4

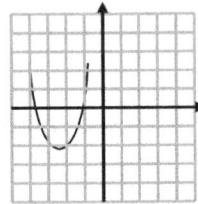

5.   maximum = 5.3
     roots = 1.7 and -1

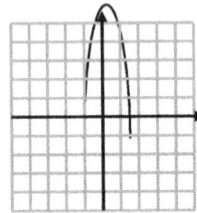

7. $f(x) = 3x^2$
     $f(-3) = 3(-3)^2 = 3(9) = 27$

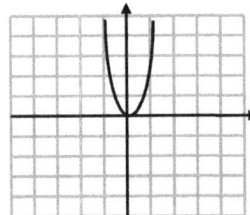

9. $f(x) = \dfrac{1}{3}x^3$
     $f(-3) = \dfrac{1}{3}(-3)^3 = \dfrac{1}{3}(-27) = -9$

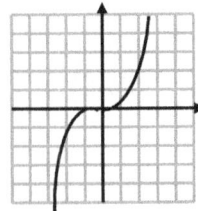

11. $V = t^2 - 14t + 48$

Substitute 3 in for V and solve for t.

$3 = t^2 - 14t + 48$ (set = 0 and factor)

$t^2 - 14t + 45 = 0$

$(t - 9)(t - 5) = 0$

$t = 9$ and $t = 5$

13. $0 = t^2 - 14t + 48$

Factor and solve for t.

$(t - 6)(t - 8) = 0$

$t = 6$ and $t = 8$

15.   When the rock hits the ground, the height is 0. Substitute 0 in for h and solve for t.

$0 = 240 - 32t - 16t^2$

$0 = -16(t^2 + 2t - 15)$

$0 = -16(t - 3)(t + 5)$

$t = 3$ and $t = -5$

Since t represents time, t cannot equal -5. Therefore the rock will hit the ground in 3 seconds.

17. Use the CALC function on the graphing calculator to determine the maximum of this function. The x-coordinate represents the time, 0.2 s. The y-coordinate represents the height, 4.6 feet.

19 a) $f(2\ sec) = -16(2)^2 + 45(2) + 400 = -16(4) + 90 + 400 = -64 + 90 + 400 = 426$ feet

b) Use the CALC function on the calculator to find the maximum y-value of the function: 431.64 ft.

c) Look at the x-intercept on the graph to calculate when the ball reaches the ground. This happens when x = 6.6 seconds.

21. $p(1.5) = 920(1.5)^2 = \$2070$

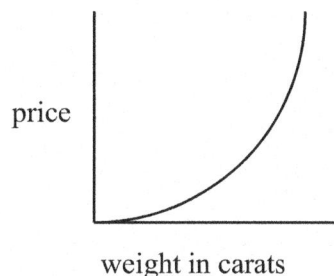

price

weight in carats

23. Substitute $a = 1$, $b = -2$ and $c = -15$ into quadratic formula and solve for x.

$$x = \frac{-(-2) \pm \sqrt{(-2)^2 - 4(1)(-15)}}{2(1)} = \frac{2 \pm \sqrt{4 + 60}}{2} = \frac{2 \pm \sqrt{64}}{2} = \frac{2 \pm 8}{2}$$

So $x = 10/2 = 5$ and $x = -6/2 = -3$

25. $3x^2 + 1 = 7x$

Set the equation equal to zero. $3x^2 - 7x + 1 = 0$

Substitute $a = 3$, $b = -7$ and $c = 1$ into quadratic formula and solve for x.

$$x = \frac{-(-7) \pm \sqrt{(-7)^2 - 4(3)(1)}}{2(3)} = \frac{7 \pm \sqrt{49 - 12}}{6} = \frac{7 \pm \sqrt{37}}{6}$$

so $x = 2.18$ and $x = 0.15$

27. $2x^2 - 3x = 9$

Set the equation equal to zero. $2x^2 - 3x - 9 = 0$

Substitute $a = 2$, $b = -3$, and $c = -9$ into the quadratic formula and solve for x.

$$x = \frac{-(-3) \pm \sqrt{(-3)^2 - 4(2)(-9)}}{2(2)} = \frac{3 \pm \sqrt{9 + 72}}{4} = \frac{3 \pm \sqrt{81}}{4} = \frac{3 \pm 9}{4}$$

$so\ x = \dfrac{12}{4} = 3\ and\ \dfrac{-6}{4} = -1.5.$

29. $h(t) = 112t - 16t^2$

Substitute 160 in for the height and solve for t. $160 = 112t - 16t^2$

$16t^2 - 112t + 160 = 0$ (set equation equal to zero)

$16(t^2 - 7t + 10) = 0$

$16(t - 5)(t - 2) = 0$

$t = 5$ seconds and 2 seconds

31. $h(t) = 560t - 16t^2$

Substitute 1056 in for height and solve for t. $1056 = 560t - 16t^2$

$16t^2 - 560t + 1056 = 0$ (set equation equal to zero)

$16(t^2 - 35t + 66) = 0$

$16(t - 33)(t - 2) = 0$

$t = 33$ seconds and 2 seconds

The rocket passes the balloon at $t = 33$ seconds on its way down.

33. $C = x^2 + 16x + 114$ (can also be done with the quadratic formula where a = 1, b= 16

            and c = 114)

Substitute 6450 in for the cost and solve for x.

$6450 = x^2 + 16x + 114$

$x^2 + 16x - 6336 = 0$ (set equal to zero and solve for x)

$(x - 72)(x + 88) = 0$

$x = 72$ and $x = -88$

Since x represents the number of units manufactured, x cannot equal a negative number.

Therefore x = 72 units.

35. $P(x) = 3x^2 + 240x - 1800$

Substitute 70,200in for the profit amount and solve for x.

$70,200 = 3x^2 + 240x - 1800$

$3x^2 + 240x - 72,000 = 0$ (set equal to zero and solve for x)

$3(x^2 + 80x - 24,000) = 0$

$3(x - 120)(x + 200) = 0$

$x = 120$ and $x = -200$

Since x represents the number of washing machines sold, x cannot equal a negative number.

Therefore x = 120 washing machines.

37. a) Neither set of data is linear. The rate of change is not consistent.

    b) They are not linear.

Practice Set 4-6

1. $f(x) = 5^x$

    $f(2) = 5^2 = 25$

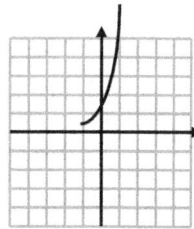

3. $f(x) = 3^x - 2$

    $f(2) = 3^2 - 2 = 9 - 2 = 7$

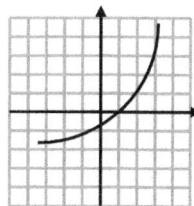

5. $f(x) = \left(\dfrac{1}{2}\right)^x$

   $f(2) = \left(\dfrac{1}{2}\right)^2 = \dfrac{1}{4}$

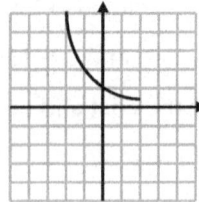

7. $f(x) = e^x$

   $f(2) = e^2 = 7.39$

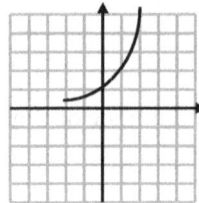

9. $f(x) = e^{-0.5x}$

   $f(2) = e^{-0.5(2)} = e^{-1} = 0.3679$

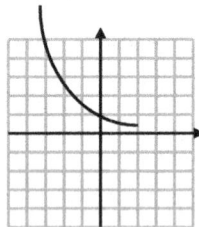

11. $f(t) = 15{,}000(1.005)^t$

    Since t = 0 corresponds to the year 2005, substitute t = 10 for the year 2015.

    $f(10) = 15{,}000(1.005)^{10} = 15{,}000(1.051140132) = 15767.10198 \approx 15{,}767$ people

13. $A = 15{,}000e^{rt}$

    $A = 15{,}000\, e^{(0.065)(10)} = 15{,}000\, e^{0.65} = 15{,}000(1.915540829) = 28733.112 \approx 28733$ people

15. $A = 12{,}500\, e^{(-0.008)(7)} = 12{,}500\, e^{-0.056} = 12{,}500(0.9455391359) = 11819.2392 \approx$ 11819 deer

17. $A = 12{,}576{,}800\, e^{(-0.0075)(10)} = 12{,}576{,}800\, e^{-0.075} = 12{,}576{,}800(0.9277434863) =$ 11,668,044.28 $\approx$ 11,668,044 people

    $A = 12{,}576{,}800\, e^{(-0.0075)(50)} = 12{,}576{,}800\, e^{-0.375} = 12{,}576{,}800(0.6872892788) =$ 8643899.801 $\approx$ 8643900 people

19. $A = 57.6\, e^{(-0.003)(17)} = 57.6\, e^{-0.051} = 57.6(0.9502786705) =$ 54.73605142 $\approx$ 54.7 million people

90

21. $y = y_0 e^{-0.04t}$

$y = 155\, e^{(-0.04)(30)} = 155\, e^{-1.2} = 155(0.3011942119) = 46.68510285 \approx 46.69$ grams

$y = 155\, e^{(-0.04)(120)} = 155\, e^{-4.8} = 155(0.008229747) = 1.275610793 \approx 1.28$ grams

23. $y = 300\, e^{-.032t}$

$y = 300\, e^{(-0.032)(10)} = 300\, e^{-0.32} = 300(0.7261490371) = 217.8447111 \approx 217.84$ grams

25. $y = A\, e^{rn}$

$y = 25{,}000\, e^{(0.045)(5)} = 25{,}000\, e^{0.225} = 25{,}000(1.252322716) = 31308.0679 =$

$31,308.07

27. $y = 2\, e^{(0.035)(5)} = 2\, e^{0.175} = 2(1.191246217) = 2.382492433 = \$2.38$

Draw the function $y = 2\, e^{0.035x}$ using your calculator and look at the table values. The value for x when y is more than 3 is 11.58 years.

Chapter 4 Review Problems

1. $T(\$12) = \$0.72$ means that you pay tax of 72 cents on a purchase of $12.

2. Answers will vary but must fit the textbook definitions.

3. Both the domain and range must be greater than or equal to 0 because purchases and taxes may be positive amounts only.

4. As the age of a person increases, his height increases up to a certain age. Then it remains steady with some shrinkage in later years.

5. As time passes, the number of bacteria in a culture increases.

6. The faster the speed of a car, the less time it takes to drive from home to school (unless you get a speeding ticket!)

7. independent variable: number of minutes; dependent variable: cost

8. domain: ≥ 0 minutes range: $\geq \$5.00$

9. $d = \dfrac{x^2}{20} + x = \dfrac{20^2}{20} + 20 = \dfrac{400}{20} + 20 = 20 + 20 = 40\,ft$

$d = \dfrac{x^2}{20} + x = \dfrac{40^2}{20} + 40 = \dfrac{1600}{20} + 40 = 80 + 40 = 120\,ft$

$d = \dfrac{x^2}{20} + x = \dfrac{60^2}{20} + 60 = \dfrac{3600}{20} + 60 = 180 + 60 = 240\,ft$

$d = \dfrac{x^2}{20} + x = \dfrac{80^2}{20} + 80 = \dfrac{6400}{20} + 80 = 320 + 80 = 400\,ft$

The graph is nonlinear with stopping distances increasing quickly.

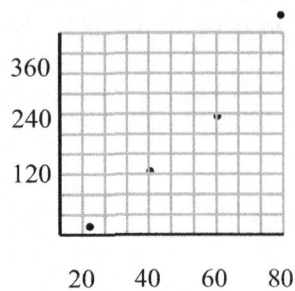

10. $f(x) = 2x - 6$

    (a) $f(-2) = 2(-2) - 6 = -10$

    (b) $f(0) = 2(0) - 6 = -6$

    (c) $f(4) = 2(4) - 6 = 2$

    (d) $f(½) = 2(½) - 6 = -5$

NOTE: for problems 11 through 15, solve for y and then replace y with $f(x)$.

11. $y = -3x + 2$ is already solved for y so, $f(x) = -3x + 2$

12. $-3y = 2x + 6$

    $y = -⅔x - 2$ (divide by -3)

    $f(x) = -⅔x - 2$

13. $x - y = 8$

    $-y = -x + 8$ (divide by -1)

    $y = x - 8$

    $f(x) = x - 8$

14. $3x + 2y = 6$

    $2y = -3x + 6$

    $y = -\frac{3}{2}x + 3$ (divide by 2)

    $f(x) = -\frac{3}{2}x + 3$

15. $4x = -y + 1$

    $y = -4x + 1$

    $f(x) = -4x + 1$

92

16. $f(x) = 2x^2 - 1$ and to fill in the chart, we will let $x = -2, -1, 0, 1$, and 2.

$f(-2) = 2(-2)^2 - 1 = 2(4) - 1 = 8 - 1 = 7$

$f(-1) = 2(-1)^2 - 1 = 2(1) - 1 = 2 - 1 = 1$

$f(0) = 2(0)^2 - 1 = 0 - 1 = -1$

$f(1) = 2(1)^2 - 1 = 2(1) - 1 = 2 - 1 = 1$

$f(2) = 2(2)^2 - 1 = 2(4) - 1 = 8 - 1 = 7$

17. Answers to a), b), and c) are found by inspection of the graph:

    a) $g(0) = -3$ b) $g(3) = 3$ c) $g(-2) = 0$

18. $C(120) = \$29 + \$0.06(120) = \$36.20$, this is the cost of renting the car and driving it 120 miles.

19. a) The base fee is $75 plus $30 times the number of credit hours, x, taken.

    $C(x) = \$75 + \$30x$

    b) $C(12) = \$75 + \$30(12) = \$75 + \$360 = \$435$

20. a) $f(x) = 1000 + 5.50x$ where x = number of books printed

    b) The domain is the set of natural numbers ≥ 250.

    c) $f(375) = 1000 + 5.50(375) = \$3062.50$

21. a) $f(x) = 1400 + 0.15(x - 14000)$ where x is taxable income

    b) $f(32,355) = 1400 + 0.15(32,355 - 14,000) = \$4153.25$

22. a) The base salary is $1000 plus 25% (0.25) times her dollars worth of sales, s

    $f(s) = \$1000 + 0.25s$

    b) $f(\$2500) = \$1000 + 0.25(\$2500) = \$1625$

23. Inspection of the table shows that when x increases by one, y increases by 2. The slope of the line is the change in y divided by the change in x, so slope = 2/1 = 2. Now use this slope and one of the x, y pairs, (2,1), from the table to find the equation of the line.

$$y = mx + b$$
$$1 = 2(2) + b$$
$$1 = 4 + b$$
$$-3 = b$$
$$f(x) = 2x - 3$$

24. Inspection of the table shows that when x increases by four, y increases by 2. The

slope of the line is the change in y divided by the change in x so, the slope is $\dfrac{2}{4} = \dfrac{1}{2}$.

Now use this slope and one of the x, y pairs, (0, 4), from the table to find the

equation of the line.

$$y = mx + b$$
$$4 = \tfrac{1}{2}(0) + b$$
$$4 = b$$
$$f(x) = \tfrac{1}{2}(x) + 4$$

25. independent variable: time; dependent variable: length of fingernails. This is a direct

variation because as times increases, the length of fingernails increases.

26. independent variable: temperature; dependent variable: time for snowman to melt.

This is an inverse variation because as temperature increases, the time it takes for the

snowman to melt decreases.

27. $s = kt$

28. $z = \dfrac{k}{p}$

29. $m = krs$

30. $x = \dfrac{k\sqrt[3]{z}}{y^2}$

31. $u = kv^3$

$$8 = k(1)^3$$

k = 8, so our working formula is: $u = 8v^3$

$$u = 8(2)^3 = 8(8) = 64$$

32. $x = \dfrac{k}{y^2}$

$$2 = \dfrac{k}{3^2} = \dfrac{k}{9}$$

k = 2(9) = 18, so our working formula is: $x = \dfrac{18}{y^2}$

$$x = \dfrac{18}{5^2} = \dfrac{18}{25}$$

$$x = 0.72$$

94

33. $s = ktg$

     $10 = k(3)(2.5)$

     $10 = 7.5k$

     $1.33 = k$ so our working formula is: $s = 1.33tg$

     $s = 1.33(7)(6) = 55.86 = 56$

34. $N = \dfrac{kL^2}{M^3}$

     $9 = \dfrac{k(2)^2}{1^3}$

     $9 = 4k$

     $2.25 = k$ so our working formula is $N = \dfrac{2.25L^2}{M^3}$

     $N = \dfrac{2.25(4)^2}{2^3} = \dfrac{2.25(16)}{8} = 2.25(2) = 4.5$

35. $d = kt^2$

     $64.4 = k(2)^2$

     $64.4 = 4k$

     $16.1 = k$ so our working formula is: $d = 16.1t^2$

     $d = 16.1(3)^2 = 16.1(9) = 144.9$ feet

36. $f = \dfrac{k}{d^2}$

     $50 = \dfrac{k}{12^2}$

     $50(144) = k$

     $7200 = k$ so our working formula is: $f = \dfrac{7200}{d^2}$

     $f = \dfrac{7200}{20^2} = \dfrac{7200}{400} = 18$ units

37. The graphs for a) and b) were done by graphing calculator and, the roots and max/mins were found with the CALC feature following the in the text.

a) minimum = -8; roots = -1, 3

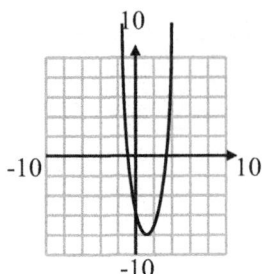

b) maximum = 9; roots = -4, 2

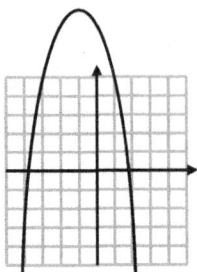

38. Graphs should be done with the graphing calculator.

(a) $f(-3) = \frac{1}{2}(-3)^4 = \frac{1}{2}(81) = 40.5$

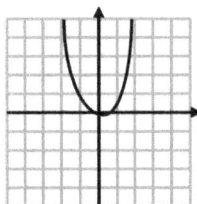

(b) $g(-3) = -2(-3)^3 = -2(-27) = 54$

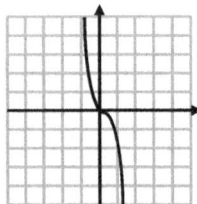

39. $x^2 - 10x - 4 = 0$

Remember the quadratic formula is: $x = \dfrac{-b \pm \sqrt{b^2 - 4ac}}{2a}$

Substitute a = 1, b = -10, and c = -4 into the formula and solve for x.

$$x = \frac{-(-10) \pm \sqrt{(-10)^2 - 4(1)(-4)}}{2(1)} = \frac{10 \pm \sqrt{100 + 16}}{2} = \frac{10 \pm \sqrt{116}}{2}$$

x = 10.385 and -.385

40. $s = 5t + 16t^2$

Substitute 74 in for s and solve for t. $74 = 5t + 16t^2$

Set equal to zero and solve for x.

$0 = 16t^2 + 5t - 74$

$0 = (16t + 37)(t - 2)$

t = -37/16 and t = 2

Since t represents time, t cannot equal a negative number.

Therefore t = 2 seconds.

41. $p = -x^2 + 16x - 24$

Substitute 40 in for p and solve for x. $40 = -x^2 + 16x - 24$

Set equal to zero and solve for x.

$x^2 - 16x + 64 = 0$

$(x - 8)(x - 8) = 0$ x = 8 telephones

42. Graphs should be done with the graphing calculator.

a) $f(-2) = 3^{-2} = \dfrac{1}{3^2} = \dfrac{1}{9}$

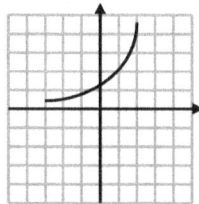

b) $g(-2) = (\tfrac{1}{2})^{-2} = 2^2 = 4$

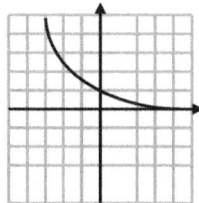

97

43. $P(10) = 100.35(1.0153)^{10} = 116.8$ million persons

44. $y = 25{,}000e^{(0.075)(24)} = 25{,}000\,e^{1.8} = 25{,}000(6.049647464) = 151241.1866$ or $151{,}242$ bacteria

45. $y = 500{,}000\,e^{(-0.05)(48)} = 500{,}000\,e^{-2.4} = 500{,}000(0.0907179533) = 45{,}358.9766$ or $45{,}359$ bacteria

46. $y = 850{,}000\,e^{(-0.02)(5)} = 850{,}000\,e^{-.1} = 850{,}000(0.904837418) = 769{,}111.8053$ or $769{,}112$ people

Chapter 4 Test

1. $25,000 was the initial value of the car (purchase price) and it decreases in value by $1500 per year after it is purchased.

2. The cost of a house is a function of its square footage. The larger the amount of square footage, the more a house usually costs. The independent variable is the square footage and the dependent variable is the cost of the house.

3. a) The independent variable is time and the dependent variable is the number of widgets produced.

   b) The initial value is 0. The function is not perfectly linear.

NOTE: For problems 4 through 6, first solve for y and then replace y with $f(x)$. Lastly find f(-4).

4. $2x - y = 4$

   $-y = -2x + 4$

   $y = 2x - 4$

   $f(x) = 2x - 4$

   $f(-4) = 2(-4) - 4 = -8 - 4 = -12$

5. $y = 5$

   $f(x) = 5$

   $f(-4) = 5$, constant

6. $3x^2 + y = -2$

$$y = -3x^2 - 2$$

$$f(x) = -3x^2 - 2$$

$$f(-4) = -3(-4)^2 - 2 = -3(16) - 2 = -50$$

7. The initial value is $22,500 and it loses value (depreciates) at the rate of $2200 each year. $V(t) = \$22,500 - \$2200t$

8. $C(12) = 1.50 + 0.75(12) = 1.50 + 9.00 = \$10.50$, which means that a cab ride of 12 miles will cost $10.50.

9. a) $f(x) = \$21 + \$28x$ where x represents the number of credit hours taken

   b) domain is the set of whole numbers 12-15 and range is the set of numbers $357, $385, $413, $441

   c) $f(14) = 21 + 28(14) = \$413$ which means that a student who registers for 14 credit hours will pay $413

10. First, find the slope by choosing two ordered pairs from the table. We chose (-1,5) and (2, -4), to put in the slope formula: $m = \dfrac{5-(-4)}{-1-2} = \dfrac{9}{-3} = -3$. Now that we have the slope, we choose one point, (-1, 5) and find the equation using $y = mx + b$.

$$5 = -3(-1) + b$$
$$5 = 3 + b$$
$$2 = b$$
$$y = -3x + 2$$
$$f(x) = -3x + 2$$

11. independent variable is time; dependent variable is the speed; The time it takes to finish a race decreases as the speed of the bike increases so this is inverse variation.

12. independent variable is time; dependent variable is the diameter of the tree; As time passes, the diameter of the trees increases so this is a direct variation.

13. $t = \dfrac{kB}{P^2}$

$$7.5 = \dfrac{k(6)}{2^2} = \dfrac{6k}{4}$$

$$7.5(4) = 6k$$

$$\dfrac{30}{6} = k$$

$5 = \text{k}$ so, our working formula is: $t = \dfrac{5B}{P^2}$

$$t = \dfrac{5(63)}{3^2} = \dfrac{315}{9} = 35$$

14. Her earnings, E, are directly proportional to the hours, H, worked. E = kH

$\$260.50 = \text{k}(8)$

$\$32.5625 = \text{k}$ is her hourly pay rate, and our working formula is: E = $32.5625H

E = $32.5625(35) = $1139.6875 = $1139.69

15. Let M = the coin's mass, T its thickness and R radius; $M = kTR^2$

$26.4 = k(0.2)(2)^2$

$26.4 = 0.8k$

$\dfrac{26.4}{0.8} = k$

$33 = k$ so our working formula is: $M = 33TR^2$

$M = 33(0.3)(4)^2 = 158.4$ grams

16. Use a graphing calculator and the CALC function as shown in the text.

maximum = 9; roots = 3, -3

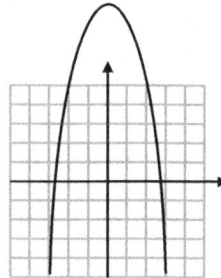

17. The graph was obtained with a graphing calculator.

$f(4) = -2^4 = -2 \cdot 2 \cdot 2 \cdot 2 = -16$

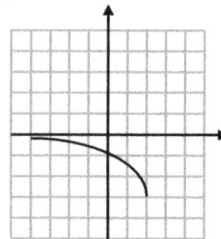

18. $C(18) = \$5200(1.045)^{18} = \$11{,}484.08959 = \$11{,}484.09$

19. $y = 260\,e^{(0.01)(50)} = 260\,e^{0.5} = 260(1.648721271) = 428.6675304 \approx 428.7$ million

20. $3x^2 - 11x = 7$ Set equal to zero and substitute into the quadratic formula.

$3x^2 - 11x - 7 = 0$ $a = 3, b = -11$ and $c = -7$

$$x = \frac{-(-11) \pm \sqrt{(-11)^2 - 4(3)(-7)}}{2(3)} = \frac{11 \pm \sqrt{121 + 84}}{6} = \frac{11 \pm \sqrt{205}}{6}$$

$x = 4.22$ and $x = -0.55$

21. $h = -16t^2 + 96t + 28$

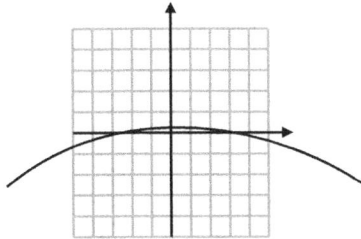

22. $-16(2)^2 + 96(2) + 28 = -16(4) + 96(2) + 28 = -64 + 192 + 28 = 156$ ft

23. $0 = -16t^2 + 96t + 28$

$a = -16, b = 96$ and $c = 28$

$$t = \frac{-96 \pm \sqrt{(96)^2 - 4(-16)(28)}}{2(-16)} = \frac{-96 \pm \sqrt{9216 + 1792}}{-32} = \frac{-96 \pm \sqrt{11008}}{-32}$$

$t = -.2787$ and $t = 6.279$ Since t represents time, t cannot equal a negative number. Therefore $t = 6.279$ seconds.

24. $y = 125,000\,e^{(0.03)(3)} = 125,000\,e^{0.09} = 125,000(1.094174284) =$

$136771.7855 \approx 136,772$ people

Chapter 5: Mathematical Models in Consumer Math

Practice Set 5-1

1. a) $(0.075)(\$25.55)=\$1.92$

   b) $(0.075)(\$155.75)=\$11.68$

   c) $(0.075)(\$256.18)=\$19.21$

3. state tax $= (0.0625)(\$25,615)=\$1600.94$

   County tax $= (0.02)(\$25,615)=\$512.30$

5. $x + (8\%)(x) = \$27,129.60$

   $1.08x = \$27,129.60$

   $x = (\$27,129.60)/(1.08) = \$25,120$ in sales

   sales tax $= \$27,129.60 - \$25,120 = \$2009.60$ or

   $(8\%)(\$25,120)=(0.08)(\$25,120)=\$2009.60$

7. $x + (7\%)(x) = \$213.95$

   $1.07x = \$213.95$

   $x = (\$213.95)/(1.07) = \$199.95$

9. Annual depreciation $= \dfrac{2150 - 500}{5} = \dfrac{1650}{5} = \$330$ per year.

11. Annual depreciation $= \dfrac{9450 - 1000}{7} = \dfrac{8450}{7} = \$1207.14$ per year.

13. $\$13,500 - \$11,880 = \$1620$ (amount depreciated)

    $\$1620/\$13,500 = 0.12 = 12\%$ depreciation

15. No pens made or sold but, fixed costs go on; a loss of $53,500

17. $(200,000)(\$1.05) - (200,000)(\$0.75) - \$53,500 =$ a profit of $6500

19. No games made or sold but, fixed costs continue; a loss of $8000

21. $(200,000)(\$5.00) - (200,000)(\$1.95) - \$8000=$ a profit of $602,000

23. $(200)(6.50) + (200)(8.95) + (200)(13.5) - (200)(2.55) - (200)(3.05) - (200)(4.15) - 3500 =$
    a profit of $340

25. (a) $C(x)=0.01x + 3500$

    (b) $C(20,500) = 0.01(20,500) + 3500 = \$3705$

    (c) $(20,500)(0.12) - 3705 =$ a loss of $1245

27. selling price $= \$35 + 55\%(\$35) = \$35 + 0.55(\$35) = \$54.25$

103

29. $48.51 - $26.95 = $21.56 markup

$21.56/$26.95 = 0.8 = 80% markup

Practice Set 5-2

1. $I = Prt$ and $M = P + I$

I = (500)(9%)(1) = $45 interest; M = 500 + 45 = $545

3. I = Prt and M = P + I

I = (1200)(12%)(1.5) = $216.00; M = 1200 + 216 = $1416.00

5. $r = \dfrac{I}{Pt} = \dfrac{120}{(500)(2)} = 0.12 = 12\%$

7. $I = Prt = (\$250)(0.18)\left(\dfrac{3}{12}\right) = \$11.25$

9. $I = Prt$ and $M = P + I$

I = $1200(12%)(0.5) = $72; Maturity Value = 1200 + 72 = $1272

payment = $1272/6 months = $212 per month

11. $I = Prt$ and $M = P + I$

I = $4500(10%)(3) = $1350; Maturity Value = 4500 + 1350 = $5850

payment = $5850/36 months = $162.50 per month

13. a) Interest = total amount paid - cash price

I = $2020 - $1600 = $420.00

b) Cost = 24($80) + $100 = $2020.00

c) $r = \dfrac{I}{Pt} = \dfrac{420}{(1500)(2)} = 0.14 = 14\%$ (remember that only $1500 was financed.)

15. 10 payments × $108.50 = $1085.00 total paid

$1085.00 - $1000 = $85.00 interest paid by financing

$r = \dfrac{I}{Pt} = \dfrac{85.00}{(1000)\left(\dfrac{10}{12}\right)} = 0.102 = 10.2\%$

17. $M = P\left(1 + \dfrac{r}{12}\right)^{12t} = 2000\left(1 + \dfrac{0.072}{12}\right)^{12} = \$2148.85$

104

19. $M = P\left(1+\dfrac{r}{4}\right)^{4t} = 25,000\left(1+\dfrac{0.06}{4}\right)^{20} = \$33,671.38$

21. a) $M = P\left(1+\dfrac{r}{4}\right)^{4t} = 900\left(1+\dfrac{0.02}{4}\right)^{12} = \$955.51$

   Interest = Maturity Value - Principal = \$955.51 - \$900 = \$55.51

   b) $M = P\left(1+\dfrac{r}{12}\right)^{12t} = 900\left(1+\dfrac{0.02}{12}\right)^{36} = \$955.61$

   Interest = Maturity Value - Principal = \$955.61 - \$900 = \$55.61

23. $M = P\left(1+\dfrac{r}{12}\right)^{12t} = 5000\left(1+\dfrac{0.04}{12}\right)^{60} = \$6104.98$;   She will not have more than \$6500.

25.   Monthly $= 10000\left(1+\dfrac{0.015}{12}\right)^{120} = \$11,617.25$

   Quarterly $= 10000\left(1+\dfrac{0.015}{4}\right)^{40} = \$11,615.08$

   Monthly - Quarterly = \$2.17

27. $A = Pe^{rt}$
   $A = \$25,000e^{(0.03)(5)} = 25000e^{0.15} = \$29.045.86$

29. $M = Pe^{rt} = 140,000e^{(0.02)(10)} = 140,000e^{0.2} = \$170,996.39$

Practice Set 5-3

1. Unpaid Balance × Monthly Interest Rate = Interest Due

   $\$350.75 \times \dfrac{0.18}{12} = 350.75 \times 0.015 = \$5.26$

3. Unpaid Balance × Monthly Interest Rate = Interest Due

   $\$655.90 \times \dfrac{0.198}{12} = 655.90 \times 0.0165 = \$10.82$

5. Monthly Interest Rate $= \dfrac{0.18}{12} = 0.015$

   June Finance Charge: $\$350.10 \times 0.015 = \$5.25$

   June Unpaid Balance: $\$350.10 + 5.25 + 25.00 + 36.75 - 75.00 = \$342.10$

   July Finance Charge: $\$342.10 \times 0.015 = \$5.13$

July Unpaid Balance: $342.10 + 5.13 + 150.40 - 200.00 = $297.63

August Finance Charge: $297.63 × 0.015 = $4.46

August Unpaid Balance: $297.63 + 4.46 + 208.75 + 55.40 - 150.00 = $416.24

7. Monthly Interest Rate = $\dfrac{0.18}{12} = 0.015$

October Finance Charge: $59.75 × 0.015 = $0.90

October Unpaid Balance: $59.75 + 0.90 + 325.88 + 34.98 - 50.00 = $371.51

November Finance Charge: $371.51 × 0.015 = $5.57

November Unpaid Balance: $371.51 + 5.57 + 59.06 + 295.60 - 150.00 = $581.74

December Finance Charge: $581.74 × 0.015 = $8.73

December Unpaid Balance: $581.74 + 8.73 + 155.67 + 230.75 - 50.00 = $926.89

9. Monthly Interest Rate = $\dfrac{0.18}{12} = 0.015$

June Fin. Chg.: 1295(0.015) = $19.43

June Unpaid Bal.: 1295 + 19.43 + 65 − 120 = $1259.43

July Fin. Chg.: 1259.43(0.015) = $18.89

July Unpaid Bal.: 1259.43 + 18.89 + 1325 + 153.85 − 350 = $2407.17

August Fin. Chg.: 2407.17(0.015) = $36.11

August Unpaid Bal.: 2407.17 + 36.11 + 415.98 + 65.09 − 575 = $2349.35

11.

Date	Balance Due	Number of Days until Balance Changed	(balance due) × (number of days)
June 3	$145.90	14	14 × $145.90 = $2042.60
June 17	$120.90	5	5 × $120.90 = $604.50
June 22	$242.85	11	11 × $242.85 = $2671.35
Total		30	$5318.45

Average Daily Balance = $\dfrac{sum\ of\ daily\ balances}{days\ in\ billing\ cycle} = \dfrac{5318.45}{30} = \$177.28$

Interest = Monthly Rate x Average Daily Balance

Interest = $\dfrac{0.18}{12} × 177.28 = 0.015 × 177.28 = \$2.66$

106

13.

Date	Balance Due	Number of Days until Balance Changed	(balance due) × (number of days)
March 1	$35.95	4	4 × 35.95 = 143.80
March 5	$0	7	7 × 0 = 0
March 12	$545.75	20	20 × 545.75 = 10,915.00
Total		31	$11,058.80

$$\text{Average Daily Balance} = \frac{sum\ of\ daily\ balances}{days\ in\ billing\ cycle} = \frac{11,058.80}{31} = \$356.74$$

Interest = Monthly Rate × Average Daily Balance

Interest = $0.015 \times 356.74 = \$5.35$

15.

Date	Balance Due	Number of Days until Balance Changed	(balance due) × (number of days)
October 5	$550.77	0	
October 5	$956.75	7	7 × $956.75 = $6697.25
October 12	$606.75	18	18 × $606.75 = $10921.50
October 30	$651.75	6	6 × $651.75= $3910.50
Total		31	$21,529.25

$$\text{Average Daily Balance} = \frac{sum\ of\ daily\ balances}{days\ in\ billing\ cycle} = \frac{21,529.25}{31} = \$694.49$$

Interest = Monthly Rate x Average Daily Balance

$$\text{Interest} = \frac{0.18}{12} \times 694.49 = 0.015 \times 694.49 = \$10.42$$

17. Finance charge per hundred = $\frac{968.05}{9500} \times 100 = 10.19$

Using the APR table on p. 176, find the closest amount on the 18-month row to $10.19. This amount is $10.19 which gives an interest rate of 12.5% APR

19. Finance charge per hundred $= \dfrac{32}{1000} \times 100 = 3.20$

Using the APR table on p. 176, find the closest amount on the 6-month row to $3.20. This amount is $3.23 which gives an interest rate of 11.0% APR.

21. Finance charge per hundred $= \dfrac{1841}{7500} \times 100 = 24.55$

Using the APR table on p. 176, find the closest amount on the 60-month row to $24.55. This amount is $24.55 which gives an interest rate of 9.0% APR.

23. Total Cost = Payments x Number of Months = $56.82 x 24 = $1363.68

Finance Charge = Total Cost - Cash Price = $1363.68 - $1250 = $113.68

Finance charge per hundred $= \dfrac{113.68}{1250} \times 100 = 9.09$

Using the APR table on p. 176, find the closest amount on the 24-month row to $9.09. This amount is $9.09 which gives an interest rate of 8.5% APR.

25. Use the APR table to find the amount per hundred that she will pay. Go to row 24 and column 5.5% to find $5.83.

Total Interest Due $= \dfrac{\text{amount financed}}{100} \times \text{amount in chart} = \dfrac{3560}{100} \times 5.83 = \$207.55$

27. First, subtract the down payment to find the amount financed.
Down Payment = 10% x 10,876 = 0.10 x 10876 = $1087.60
Amount financed = Cash Price - Down Payment = $10,876 - $1087.60 = $9788.40

Use the APR table to find the amount per hundred that she will pay. Go to row 48 and column 7.5% to find $16.06.

Total Interest Due $= \dfrac{\text{amount financed}}{100} \times \text{amount in chart} = \dfrac{9788.40}{100} \times \$16.06 = \$1572.02$

Needed to pay the loan = balance due + interest = $9788.40 + $1572.02 = $11,360.42

Monthly Payments $= \dfrac{11360.42}{48} = \$236.68/\text{month}$

29. First, subtract the down payment to find the amount financed.
Amount financed = Cash Price - Down Payment = $4578 - $500 = $4078
Use the APR table to find the amount per hundred that she will pay. Go to row 36 and column 4.5% to find $7.09.

108

$$\text{Total Interest Due} = \frac{\text{amount financed}}{100} \times \text{amount in chart} = \frac{4078}{100} \times 7.09 = \$289.14$$

Needed to pay the loan = balance due + interest = 4078 + 289.14 = \$4367.14

$$\text{Monthly Payments} = \frac{4367.14}{36} = \$121.31/\text{month}$$

Practice Set 5-4

1. Simple interest is calculated on the full amount borrowed for the full term of the loan. Then interest plus principal are added and divided into equal monthly payments.

3. lower payments, repairs under warranty, (many other possible answers)

5. you may sell the car at any time, you may customize it, (other answers are possible)

7. $46,090 + 5\%(46,090) = \$48,394.50$

9. $29,460 + 6\%(29,460) = \$31,227.60$

11. $56,904 + 7\%(56,904) = \$60,887.28$

13. $I = Prt = (37,884)(0.029)(5) = \$5493.18$

   Pmt = (37,884.00 + 5493.18)/60 = \$722.96 monthly

15. $I = Prt = (15,747)(0.05)(3) = \$2362.05$

   Pmt = (15,747.00 + 2362.05)/36 = \$503.03 monthly

17. $I = Prt = (50,187)(0.019)(6) = \$5721.32$

   Pmt = (50,187 + 5721.32)/72 = \$776.51 monthly

19. cost = (317.50)(36) = \$11,430.00

   Interest = 11,430 − 9000 = \$2430.00

21. cost = 3290 + (339.97)(48) = \$19,608.56

   Interest = 19,608.56 − 16,450 = \$3158.56

23. $I = 36(317.50) − 9000 = \$2430.00$

   $$r = \frac{I}{Pt} = \frac{2430}{9000(3)} = 0.09 = 9\%$$

25. $I = 48(339.97) − (16450 − 3290) = \$3158.56$

   $$r = \frac{I}{Pt} = \frac{3158.56}{13160(4)} = 0.06 = 6\%$$

27. $I = 500 + 127.20(36) - 4500 = 579.20$

$$\frac{579.20}{4000} \times 100 = 14.48 \; ; \text{ find closest to 14.48 on row 36 of the APR table}$$

APR = 9.0%

29. APR table row 36, 4.5% = \$7.09 per \$100 financed

Amt. Financed = 8578 − 500 = \$8078

$$I = 7.09\left(\frac{8078}{100}\right) = \$572.73$$

Pmt = (8078 + 572.73)/36 = \$240.30

31. $A = 426.71 + 48(426.71) + 895 = \$21,803.79$

33. $25,000 - 10\%(25,000) = \$22,500$

$$\text{Pmt} = \frac{P+I}{48} = \frac{P + Prt}{48} = \frac{22,500 + 22,500 \times .029 \times 4}{48} = \$523.13$$

Total cost = 2500 + 48(523.13) = \$27,610

35. Extra miles = 65,200 − 4(15,000) = 5200 miles

5200(0.18) = \$936.00 extra

Practice Set 5-5

1. interest rate, fees, points, type of loan, etc.

3. subtract your monthly bills from your gross income and multiply the balance by 36%

5. use the maximum amount of loan formula and the amortization table (page 189)

7. down pmt = 10%(153,250) = \$15,325; amt. fin = 153,250 − 15,325 = \$137,925

9. down pmt = 25%(179,900) = \$44,975; amt. fin = 179,900 − 44,975 = \$134,925

11. down pmt = 30%(315,500) = \$94,650; amt. fin = 315,500 − 94,650 = \$220,850

13. (gross income − monthly bills)(36%) = (985.65 − 135.50)(36%) = \$306.05

15. (2200 − 550)(36%) = \$594.00

17. (3700 − 650)(36%) = \$1098.00

19. $P = \$185,000\left[\dfrac{\dfrac{0.035}{12}\left(1 + \dfrac{0.035}{12}\right)^{12 \times 25}}{\left(1 + \dfrac{0.035}{12}\right)^{12 \times 25} - 1}\right] = \$926.16$

21. $P = \$110{,}500\left[\dfrac{\dfrac{0.045}{12}\left(1+\dfrac{0.045}{12}\right)^{12\times30}}{\left(1+\dfrac{0.045}{12}\right)^{12\times30}-1}\right] = \$559.89$

23. $P = \$113{,}650\left[\dfrac{\dfrac{0.0585}{12}\left(1+\dfrac{0.0585}{12}\right)^{12\times20}}{\left(1+\dfrac{0.0585}{12}\right)^{12\times20}-1}\right] = \$804.42$

25. $P = \dfrac{85{,}000}{1000}\times 5.01 = \$425.85$

27. $P = \dfrac{112{,}000}{1000}\times 7.65 = \$856.80$

29. $P = \dfrac{265{,}000}{1000}\times 5.37 = \$1423.05$

31. Max. Loan $= \dfrac{750}{4.49}\times 1000 = \$167{,}038$

33. Max. Loan $= \dfrac{1850}{5.01}\times 1000 = \$369{,}262$

35. Max. Loan $= \dfrac{635}{5.07}\times 1000 = \$125{,}247$

37. down pmt = 25%(250,000) = $62,500; amt. fin. = 250,000 − 62,500 = $187,500

$P = \$187{,}500\left[\dfrac{\dfrac{0.0575}{12}\left(1+\dfrac{0.0575}{12}\right)^{12\times30}}{\left(1+\dfrac{0.0575}{12}\right)^{12\times30}-1}\right] = \$1094.20$

39. a) down pmt. = 15%(118,500) = $17,775; balance = $100,725

$P = \$110{,}725\left[\dfrac{\dfrac{0.0425}{12}\left(1+\dfrac{0.0425}{12}\right)^{12\times30}}{\left(1+\dfrac{0.0425}{12}\right)^{12\times30}-1}\right] = \$495.51$

   b) A = 360(495.51) + 17,775 = $196,158.60
   c) go to www.fha.com

41. go to www.fha.com

Practice Set 5-6

1. An economical way of helping an individual deal with potential severe financial losses by pooling money with a large group of people.

3. The total amount of insurance specified by the insurance policy purchased is the face value.

5. The payment made to the policyholder in the event of an insured loss is called an indemnity.

7. $20,000 for bodily injury per person injured, $40,000 bodily injury per accident, and $15,000 property damage per accident (20/40/15)

9. Collision insurance pays for repairs to your own car if you are responsible for an accident.

11. $101 + 99 = 200; 200(1.50) = \$300$

13. $91 + 95 = 186; 186(1.10) = \$204.60$

15. $136 + 107 = 243; 243(2.60) = \$631.80$

17. $57 + 94 = 151; 151(1.40) = \$211.40$

19. $52 + 73 = 125; 125(1.40) = \$175.00$

21. $64 + 125 = 189; 189(1.40) = \$264.60$

23. $(86 + 83 + 55 + 82)(1.90) = \$581.40$

25. a) 8000

    b) $0

Practice Set 5-7

1. a) $71.61(1500) = \$114,915$

    b) $0.02(114,915) = \$2298.30$

3. a) $23.90(50) = \$1195.00$

    b) $0.015(1195) + 0.10(50) = \$22.93$

5. a) $37.80(1100) = \$41,580$

    b) $0.021(41,580) = \$873.18$

7. a) $30.02(40) = \$1200.80$

    b) $0.025(1200.80) + 0.125(40) = \$35.02$

9. a) $500(0.40) = \$200$

    b) $291,200(100) = 29,120,000$

    c) $11.11

    d) $11.11 + 0.18 = \$11.29$

    e) $-0.18/11.29 = -1.59\%$, a decrease of 1.59%

112

11. a) $50(0.12) = \$6.00$

   b) $75{,}750(100) = 7{,}750{,}000$

   c) $\$25.97$

   d) $25.97 - 0.33 = \$25.64$

   e) $0.33/25.04 = 1.29\%$ increase

13. $2155 \div 8.62 = 250$ shares

15. $10{,}150 \div 29 = 350$ shares

17. Bonds are issued by companies and governments to raise money for projects. Bonds are generally safer investments than stocks.

19. Invest more of your funds in stocks and mutual funds since, over long periods of time, these types of investments have a history of higher growth rates than bonds.

Practice Set 5-8

1. $P = (\$15670)(9.5\%) = (\$15670)(0.095) = \$1488.65$

3. $P = \$1245 + (3.5\%)(\$65000) = \$1245 + \$2275 = \$3520$

5. $P = 2200 + (2\%)(12350 - 7500) = 2200 + (0.02)(4850) = 2200 + 97 = \$2297$

7. $P = (2.5\%)(125{,}000) = (0.025)(125{,}000) = \$3125$

9. Tip $= 15\%(65) = 0.15(65) = \$9.75$

11. $0.06(150{,}000) = 9000; \ 0.60(9000) = \$5400$ commission

13. $425/5000 = 0.085 = 8.5\%$

15. a) tax $= 1275 + 7\%(48745 - 21250) = \$3199.65$

    b) tax $= 765 + 7\%(48745 - 12750) = \$3284.65$

17. Fixed rate tax: $4.5\%(31{,}000) = 0.045(31{,}000) = \$1395$

    Tax Table: $765 + 0.07(22{,}450 - 12{,}750) = 765 + 0.07(9700) = \$1444$

    He owes more if he uses the tax table.

19. Income $= 0.045(12)(50{,}000) = \$27{,}000$

    Tax $= 765 + 0.07(16{,}750 - 12{,}750) = \$1045$

21. a) $C = 800 + 9.5\%(S) = 800 + 0.095S$

    b) The above equation is written in slope-intercept form. slope $= m = 9.5\% = 0.095$ dollars per sale

c) Slope units would be dollars of commission earned per dollars of sales made

d) Yes because there is only one commission for each amount of sales made.

e) "Commission" would go on the y-axis since it is dependent upon the amount of sales.

f) $2525 = 800 + 9.5\%(S)$

$2525 - 800 = 0.095S$

$1725 = 0.095S$

$S = \$18157.89$ (rounded to the nearest cent)

23. Housing   Rent $590
           Utilities   25
           Phone   45
Loans   Car   125
           Food   320
   Entertainment   150
           TV   45
     Gasoline   150
Total Budget Amt. = $1450

Savings desired = $0.07(1855) = \$129.85$

The total budgeted plus the desired savings amount is less than the total take-home pay,

so saving 7% is possible.

25. $0.15(1978) = \$296.70$

27. a) 36%(income – expenses) = estimated maximum loan payment

$0.36(4250 - 215 - 45 - 55 - 45) = \$1400.40$

b) taxes = $0.15(4250) = 637.50$

utilities = $0.0425(4250) = 180.63$

food = $0.135(4250) = 573.75$

savings = $0.055(4250) = 233.75$

entertainment = $0.04(4250) = 170.00$

gasoline/car = $0.05(4250) = 212.50$

insurance = $0.02(4250) = 85.00$

Total Budget Amount = $3853.13 (including house payment and monthly expenses)

This leaves $396.47 per month out of the gross pay.

29. Answers will vary.

Chapter 5 Review Problems

1. County tax = 0.025(12,548) = $313.70

   State tax = 0.045(12,548) = $564.66

2. $x + 8.5\%(x) = 4285.75$

   $1.085x = 4285.75$

   $x = \$3950$

3. $0.05(12.80) + 0.06(26.50) = \$2.23$

4. tip = $15 - 12.50 = \$2.50$; $2.50/12.50 = 0.2 = 20\%$

5. straight line depreciation amt. = $(25,000 - 1500)/10 = \$2,350$

6. Profit = gross sales - costs = $75(275) - [1250 + 75(125)] = \$10,000.00$

7. $(200\%)(475) = 2(475) = \$950$ markup; selling price = $950 + 475 = \$1,425$

8. Commission only: $25\%(5000) = \$1250.00$

   Salary plus commission: $1000 + 5\%(5000) = 1000 + 250 = \$1250.00$

   So for July, Julian would make the same either way.

9. $I = Prt$; $r = \dfrac{I}{Pt} = \dfrac{162.50}{(5000)(0.5)} = 0.065 = 6.5\%$ interest rate

10. part 1: $I = Prt = 15000(5.5\%)(2) = \$1650$ and M = P + I = 15000 + 1650 = $16,650

    part 2: payments = $16,650/(2 \cdot 12) = 16,650/24 = \$693.75$ per month

11. $I = Prt = 1000(3\%)(6) = 1000(0.03)(6) = \$180$

12. $r = 5.8\%$ and $t = 2$

$$M = 8000\left(1 + \frac{0.058}{4}\right)^{4(2)} = 8000(1.122060854) = 8976.4868... = \$8976.49$$

13. Find the compound amount, M, and subtract the principal, P, to get the interest that was added.

$$I = M - P = 2500\left(1 + \frac{0.035}{12}\right)^{12 \times 5} - 2500 = 2977.35 - 2500 = \$477.35$$

14. $A = 10,000e^{(0.04)(20)} = 22,255.40928 = \$22,255.41$

15. $3150.75\left(\dfrac{0.15}{12}\right) = \$39.39$

16.

Date	Balance Due	Number of Days until Balance Changed	(balance due) × (number of days)
June 1	$1415.90	14	14 × $1415.90 = 19,822.60
June 15	$1290.90	8	8 × $1290.90 = 10,327.20
June 23	$1312.85	8	8 × 1312.85 = 10,502.80
Total		30	$40,652.60

$$\text{Average Daily Balance} = \frac{\text{sum of daily balances}}{\text{days in billing cycle}} = \frac{40,652.60}{30} = \$1355.09$$

Interest = Monthly Rate x Average Daily Balance

$$\text{Interest} = \frac{0.18}{12} \times 1355.09 = 0.015 \times 1355.09 = \$20.33$$

17. APR table row 36, 8% = $12.81 per $100 financed

Amt. Financed = 2350 − 500 = $1850

$$I = 12.81\left(\frac{1850}{100}\right) = \$236.99$$

Pmt = (1850 + 236.99)/36 = $57.98

18. 26,998 + 0.05(26,998) = $28,347.90

19. $I = Prt$ = 27,885(0.009)(5) = $1254.83

Pmt = (27,885 + 1254.83)/60 = $485.67

20. cost = 36(196.80) = $7084.80

I = 7084.80 − 6500 = $584.80

$$r = \frac{584.80}{6500(3)} = 0.0299\ldots = 3\%$$

21. 395.59 + 849 + 48(395.59) = $20,232.91

22. 62,148 − 4(15,000) = 2148 miles

Extra cost = (2148 miles)($0.15/mile) = $322.20

23. Cost = 25,000 + 25,000(0.036)(5) = $29,500

Pmt = 29,500/60 = $491.67 monthly

24. 36%(income − expenses) = estimated maximum loan payment

$$0.36(4988 - 500.30) = \$1615.57$$

25. Balance to finance = 145000-(20%)(145000)=$116,000

$$P = 116000\left(\frac{\frac{0.0625}{12}\left(1+\frac{0.0625}{12}\right)^{12*30}}{\left(1+\frac{0.0625}{12}\right)^{12*30}-1}\right) = 116000\left(\frac{0.0337977416}{5.48916638}\right) = \$714.24$$

26. $P = \dfrac{148,500}{1000} \times 5.68 = \$843.48$

27. (64 + 104)(1.40) = $235.20

28. (129 + 108)(1.40) = $331.80

29. $20,000

30. 150(395.31) + 10.99 = $59,307.49

31. a) 450(27.28) = $12,276

    b) 0.025(12,276) = $306.90 commission

32. 87250/34.90 = 2500 shares

33. x(12500) = 812.50

    x = 812.50/12500 = 0.065 = 6.5% commission

34. Salary = 400 + 6.5%(6000-1200) = 400 + 0.065(4800) = $712.00

35. T = $1275 + 0.07(x - 21250)

36. T = 1275 + 7%(22500-21250) = 1275 + 87.5 = $1362.50

37. 0.15(2386) = $357.90

Chapter 5 Test

1. Tax = 0.035(24,500) = $857.50

2. 12.40 + 1.25(12.40) = $27.90

3. Depreciation amount = (18,000 - 2500)/10 = $1550.00 per year

4. P = 14,500(2.20) - [3575 + 14500(0.82)] = $16,435.00

5. P = 200 + 0.12(1275) = $353.00

6. T = 18%(55,250) = 0.18(55,250) = $9945.00

7. $M = 2500\left(1+\dfrac{0.045}{4}\right)^{4(15)} = 2500(1.956645179) = 4891.61$

117

8. $250/500 = 0.05 = 5\%$

9. $M = 7500e^{(0.055)(6)} = 10432.26096 = \$10,432.26$

10. $5250.05\left(\dfrac{0.08}{12}\right) = \$35.00$

11.

Date	Balance Due	Number of Days until Balance Changed	(balance due) ×(number of days)
July 1	$115.90	14	$14 \times 115.90 = 1622.60$
July 15	$90.90	5	$5 \times 90.90 = 454.50$
July 20	$316.45	12	$12 \times 316.45 = 3797.40$
Total		31	$5874.50

$$\text{Average Daily Balance} = \frac{sum\ of\ daily\ balances}{days\ in\ billing\ cycle} = \frac{5874.50}{31} = \$189.50$$

Interest = Monthly Rate × Average Daily Balance

$$\text{Interest} = \frac{0.18}{12} \times 189.50 = 0.015 \times 189.50 = \$2..84$$

12. Amount to finance $= 1385 - 100 = \$1285$

$$P = 5.29\left(\frac{1285}{100}\right) = \$67.98$$

Pmt $= (1285 + 67.98)/24 = \$56.38$

13. $I = Prt = 57,885(0.009)(6) = \$3125.79$

Pmt. $= (57,885 + 3125.79)/72 = \$847.38$

14. Cost $= 571 + 54(571) + 795 = \$32,200$

15. Amount to finance $= 36,200 - 600 = \$35,600$

I = Prt $= 35,600(0.019)(6) = \$4058.40$

Pmt. $= (35,600 + 4058.40)/72 = \$550.82$

16. Amount to finance $= 375,000 - 0.25(375,000) = \$281,250$

Table 4.4, row 5.5%, 30 years, $= 5.68$

$$P = \frac{281,250}{1000} \times 5.68 = \$1597.50$$

17. $(103 + 101)(1.50) = \$306$

118

18. $0

19. $175(49.5) + 250(29.88) + 2(9.99) = \$16{,}152.48$

20. a) $50(5.15) = \$257.50$

   b) Commission $= 0.015(257.50) + 0.15(50) = \$11.37$

Chapter 6: Modeling with Systems of Equations

Practice Set 6-1

1. (3, -2) solves both equations: $3(3) + 4(-2) = 1$ and $3 - (-2) = 5$

3. (-3, -1) solves both equations: $-3 - 4(-1) = 1$ and $5(-3) - 3(-1) = -12$

5. $\left(\dfrac{1}{2}, 1\right)$ solves both equations: $2(\dfrac{1}{2}) + 1 = 2$ and $6(\dfrac{1}{2}) - 4(1) = -1$

Note: for numbers 7-25, small graphs here would be of little use here to show the actual crossing points in enough detail to be readable

7. (1, 2) by graphing in the TI-83 plus or TI-84

9. (2, 0) by graphing in the TI-83 plus or TI-84

11. (4, 1) by graphing in the TI-83 plus or TI-84

13. (0, 1) by graphing in the TI-83 plus or TI-84

15. infinite solutions - $\{(x,y)| x + 6y = 27\}$; by graphing in the TI-83 plus or TI-84

17. no solution - parallel lines; by graphing in the TI-83 plus or TI-84

19. (-2, -4) by graphing in the TI-83 plus or TI-84

21. (-3, -1) by graphing in the TI-83 plus or TI-84

23. infinite solutions – $\{(x,y)| x + y = 4\}$; by graphing in the TI-83 plus or TI-84

25. no solution - parallel lines; by graphing in the TI-83 plus or TI-84

27. (2, 0) by graphing in the TI-83 plus or TI-84

29. (2, -1) by graphing in the TI-83 plus or TI-84

31. infinite solutions – $\{(x,y)| 2x - 3y = 1\}$; by graphing in the TI-83 plus or TI-84

Practice Set 6-2

1. $x = 2y$; Substitute this for x in the second equation.

$$2y + y = 18$$
$$3y = 18$$
$$y = 6$$

121

Substitute y = 6 into the first equation.

$$x = 2y = 2(6) = 12 \qquad \text{The solution is (12, 6)}$$

3. x = 5y + 5; Substitute this in for x in the first equation.

$$5(5y + 5) + 3y = 11$$

$$28y + 25 = 11$$

$$28y = -14$$

$$y = -0.5$$

Substitute y = 0.5 into the second equation.

$$x = 5y + 5 = 5(-0.5) + 5 = 2.5 \qquad \text{The solution is (2.5, - 0.5)}.$$

5. Solve the first equation for y: y = 3x - 4 and substitute this in for y in the second equation.

$$9x - 5(3x - 4) = -10$$

$$9x - 15x + 20 = -10$$

$$-6x = -30$$

$$x = 5$$

Substitute x = 5 into the first equation.

$$y = 3(5) - 4 = 15 - 4 = 11 \qquad \text{The solution is (5, 11)}.$$

7. 2x - 3y = 7

4x + 3y = 5   Multiply equation #1 by -2 and add to equation #2.

$$-4x + 6y = -14$$

$$\underline{4x + 3y = \quad 5}$$

$$9y = \quad -9$$

$$y = \quad -1$$

Substitute y = -1 into the first equation: 2x -3(-1) = 7

$$2x + 3 = 7$$

$$2x = 4$$

$$x = 2 \qquad \text{The solution is: (2, -1)}.$$

122

9. $x + 7y = 12$

    $3x - 5y = 10$ Multiply equation #1 by -3 and add to equation #2.

            $-3x - 21y = -36$

            $\underline{3x\ -\ 5y\ =\ 10}$

              $-26y = -26$

                $y = 1$

    Substitute $y = 1$ into the first equation: $x + 7(1) = 12$

                                        $x + 7 = 12$

                                          $x = 5$ The solution is: (5, 1).

11. $x - y = 3$ (after re-arrangement)

    $x - 3y = -2$ (after re-arrangement)

    Multiply equation #1 by -1 and add to equation #2.

          $-x + y\ = -3$

          $\underline{x - 3y = -2}$

            $-2y\ = -5$

              $y = 2.5$

    Substitute $y = 2.5$ into the first equation: $x - 2.5 = 3$

                                       $x = 5.5$ The solution is (5.5, 2.5).

13. $3x - 2y = 6$

    $-6x + 4y = -12$ Multiply equation #1 by 2 and add to equation #2.

            $6x - 4y = 12$

           $\underline{-6x + 4y = -12}$

              $0 = 0$ this means that these are the same line

           The solution is infinite number of solutions or $\{(x,y)|\ 3x - 2y = 6\}$.

15. multiplying diagonals gives $(1)(5) - (2)(2) = 5 - 4 = 1$

17. multiplying diagonals gives $(3)(4) - (6)(2) = 12 - 12 = 0$

19. multiplying diagonals gives $(3)(4) - (-2)(5) = 12 + 10 = 22$

21. $x = \dfrac{\begin{vmatrix} 1 & -4 \\ 13 & 3 \end{vmatrix}}{\begin{vmatrix} 1 & -4 \\ 2 & 3 \end{vmatrix}} = \dfrac{3 - (-52)}{3 - (-8)} = \dfrac{55}{11} = 5$

Now substitute into equation #1. 5 - 4y = 1

$\qquad\qquad\qquad\qquad$ -4y = -4

$\qquad\qquad\qquad\qquad\qquad$ y = 1 $\qquad\qquad\qquad$ The solution is (5, 1).

23. $x = \dfrac{\begin{vmatrix} 36 & -2 \\ 47 & 4 \end{vmatrix}}{\begin{vmatrix} 6 & -2 \\ 5 & 4 \end{vmatrix}} = \dfrac{144 - (-94)}{24 - (-10)} = \dfrac{238}{34} = 7$

Now substitute into equation #1.

$\qquad\qquad\qquad$ 6(7) - 2y = 36

$\qquad\qquad\qquad$ 42 - 2y = 36

$\qquad\qquad\qquad\qquad$ -2y = -6

$\qquad\qquad\qquad\qquad\quad$ y = 3 $\qquad\qquad\qquad$ The solution is (7, 3).

25. $x = \dfrac{\begin{vmatrix} 28 & -7 \\ -20 & 5 \end{vmatrix}}{\begin{vmatrix} 6 & -7 \\ -4 & 5 \end{vmatrix}} = \dfrac{140 - 140}{30 - 28} = \dfrac{0}{2} = 0$

Now substitute into equation #1.

$\qquad\qquad\qquad$ 6(0) - 7y = 28

$\qquad\qquad\qquad\qquad\quad$ y = -4 $\qquad\qquad\qquad$ The solution is (0, -4).

27. $x = \dfrac{\begin{vmatrix} -10 & -1 \\ 4 & 1 \end{vmatrix}}{\begin{vmatrix} 2 & -1 \\ 1 & 1 \end{vmatrix}} = \dfrac{-10 - (-4)}{2 - (-1)} = \dfrac{-6}{3} = -2$

Now substitute into equation #2.

$\qquad\qquad\qquad$ -2 + y = 4

$\qquad\qquad\qquad\qquad$ y = 6 $\qquad\qquad\qquad$ The solution is (-2, 6).

29. $x = \dfrac{\begin{vmatrix} 13 & -2 \\ -26 & 4 \end{vmatrix}}{\begin{vmatrix} 3 & -2 \\ -6 & 4 \end{vmatrix}} = \dfrac{52-52}{12-12} = \dfrac{0}{0} =$ infinite solutions or $\{(x,y) \mid 3x - 2y = 13\}$

31. $x = \dfrac{\begin{vmatrix} 0 & 1 \\ 86 & -15 \end{vmatrix}}{\begin{vmatrix} -3 & 1 \\ 2 & -15 \end{vmatrix}} = \dfrac{0-86}{45-2} = \dfrac{-86}{43} = -2$

Now substitute into equation #1.

$$y = 3(-2)$$
$$y = -6$$

The solution is (-2, -6).

33. $x = \dfrac{\begin{vmatrix} 10 & -3 \\ 2 & 5 \end{vmatrix}}{\begin{vmatrix} 2 & -3 \\ 2 & 5 \end{vmatrix}} = \dfrac{50-(-6)}{10-(-6)} = \dfrac{56}{16} = 3.5$

Now substitute into equation #1.

$$2(3.5)-3y = 10$$
$$7 - 3y = 10$$
$$-3y = 3$$
$$y = -1$$

The solution is (3.5, -1).

35. $x = \dfrac{\begin{vmatrix} 0 & 4 \\ 17 & -5 \end{vmatrix}}{\begin{vmatrix} 1 & 4 \\ 3 & -5 \end{vmatrix}} = \dfrac{0-68}{-5-12} = \dfrac{-68}{-17} = 4$

Now substitute into equation #1.

$$4 = -4y$$
$$y = -1$$

The solution is (4, -1).

125

37. $x = \dfrac{\begin{vmatrix} -36 & 3 \\ 22 & -5 \end{vmatrix}}{\begin{vmatrix} 2 & 3 \\ 3 & -5 \end{vmatrix}} = \dfrac{180 - 66}{-10 - 9} = \dfrac{114}{-19} = -6$

Now substitute into equation #1.

$$3y + 2(-6) = -36$$
$$3y - 12 = -36$$
$$3y = -24$$
$$y = -8$$

The solution is (-6, -8).

39. $x = \dfrac{\begin{vmatrix} 2.2 & -0.4 \\ 1.6 & -1 \end{vmatrix}}{\begin{vmatrix} 0.2 & -0.4 \\ 0.5 & -1 \end{vmatrix}} = \dfrac{-2.2 - (-0.64)}{-0.2 - (-0.2)} = \dfrac{-1.56}{0} = $ No solution; parallel lines

41. $x = \dfrac{\begin{vmatrix} 1 & 3 \\ 2 & 2 \end{vmatrix}}{\begin{vmatrix} 2 & 3 \\ 3 & 2 \end{vmatrix}} = \dfrac{2 - 6}{4 - 9} = \dfrac{-4}{-5} = \dfrac{4}{5}$

Now substitute into equation #1.

$$2\left(\dfrac{4}{5}\right) + 3y = 1$$
$$\dfrac{8}{5} + 3y = 1$$
$$3y = -\dfrac{3}{5}$$
$$y = -\dfrac{1}{5}$$

The solution is $\left(\dfrac{4}{5}, -\dfrac{1}{5}\right)$.

43. $x = \dfrac{\begin{vmatrix} 1 & -2/5 \\ 2 & -3/2 \end{vmatrix}}{\begin{vmatrix} 2/3 & -2/5 \\ 5/2 & -3/2 \end{vmatrix}} = \dfrac{-3/2 - (-4/5)}{-6/6 - (-10/10)} = \dfrac{-7/10}{0} = $ No solution: parallel lines

45. By substitution, solve equation #1 for y: $y = 2x - 2$

Substitute the expression for y into equation #2.

$$4x + 3(2x - 2) = 24$$
$$4x + 6x - 6 = 24$$
$$10x = 30$$
$$x = 3$$

Now substitute this result into the original equation.

$$y = 2x - 2 = 2(3) - 2 = 4$$ The solution is (3, 4).

47. Substitute for y from equation #1 into equation #2.

$$2x - 1 = x + 1$$
$$x = 2$$

Now use this value in equation #2.

$$y = 2(2) - 1 = 3$$ The solution is (2, 3).

49. Solve by addition. Eliminate y by multiplying the first equation by 5 and the second equation by -2 and then adding the two equations together.

$$25x - 10y = 70$$
$$\underline{-12x + 10y = -18}$$
$$13x = 52$$
$$x = 4$$

Substitute this result into the first equation.

$$5(4) - 2y = 14$$
$$20 - 2y = 14$$
$$-2y = -6$$
$$y = 3$$ The solution is (4, 3).

51. Solve by substitution. Rearrange the first equation to give: $y = x - 8$.

Now substitute $x - 8$ into the second equation.

$$3(x - 8) = x + 2$$
$$3x - 24 = x + 2$$
$$2x = 26$$
$$x = 13$$

Now substitute x = 13 into the first equation.

$$y = x - 8 = 13 - 8 = 5$$ The solution is (13, 5).

53. Solve by substitution. Divide equation #1 by -2; -7x + 3 = y or y = -7x + 3

Substitute into equation #2.

$$7x + (-7x + 3) = 3$$
$$0 = 0$$

The lines are the same line; infinite solutions or $\{(x,y) \mid 7x + y = 3$.

55. Solve by addition. Multiply equation one by 4 and equation two by 3 and then add.

$$28x + 12y = -32$$
$$\underline{15x - 12y = -183}$$
$$43x = -215$$
$$x = -5$$

Substitute this result into equation one. $7(-5) + 3y = -8$
$$-35 + 3y = -8$$
$$3y = 27$$
$$y = 9 \qquad \text{The solution is (-5, 9)}$$

57. No, because (2,3) is not a solution for equation one. $3(2) + 5 = 3; 6 + 5 \neq 3$.

59. x + y = 20

$\underline{x - y = 2}$ Solve by addition .
$$2x = 22$$
$$x = 11$$

Substitute this result into equation one. 11 + y = 20
$$y = 9 \qquad \text{The solution is (11, 9)}.$$

61. Set up two equations as follows:

$$x + y = 30$$
$$2x = 3y$$

Solve equation one for x: x = -y + 30.

Solve by substituting this expression into equation two.

$$2(-y + 30) = 3y$$
$$-2y + 60 = 3y$$
$$5y = 60$$
$$y = 12$$

128

Substitute this result into equation one.

$$x = -12 + 30$$
$$x = 18$$

The solution is (18, 12).

63. Set up two equations: $\dfrac{x + y}{2} = x - y$ and $x + y = 64$

Solve equation two for x: $x = 64 - y$

Rearrange equation one. $x + y = 2x - 2y$. Now substitute the expression for *x* from equation one into equation two.

$$64 - y + y = 2(64 - y) - 2y$$
$$64 = 128 - 4y$$
$$4y = 64$$
$$y = 16$$

Substitute this result into equation two.

$$x = 64 - y = 64 - 16 = 48$$

The solution is (48, 16).

65. Set up two equations: $3x = 4y$ and $x + y = 46$

Solve equation two for *x:* $x = 46 - y$

Substitute this expression into equation one.

$$3(46 - y) = 4y$$
$$138 - 3y = 4y$$
$$138 = 7y$$
$$y = \frac{138}{7}$$

Substitute this result into equation two.

$$x = 46 - \frac{138}{7} = \frac{184}{7}$$

The solution is $\left(\dfrac{184}{7}, \dfrac{138}{7} \right)$.

Practice Set 6-3

1. Let x = one-bedroom apartments and y = two-bedroom apartments.

Set up two equations: $x + y = 20$ and $\$650x + \$825y = \$14,400$

Solve equation #1 for *y* and substitute into equation #2: $y = 20 - x$

$$650x + 825(20 - x) = 14400$$
$$650x + 16500 - 825x = 14400$$
$$-175x = -2100$$
$$x = 12$$

There are 12 one-bedroom units and 8 two-bedroom units.

129

3. Let x = double rooms and y = single rooms.

$$90x + 80y = 6930$$
$$x + y = 80$$

Solve equation #2 for x and substitute into equation #1: $x = 80 - y$

$$90(80 - y) + 80y = 6930$$
$$7200 - 90y + 80y = 6930$$
$$7200 - 10y = 6930$$
$$-10y = -270$$
$$y = 27$$

Find x: $x = 80 - y = 80 - 27 = 53$

There are 53 double rooms and 27 single rooms.

5. Let x = amount at 5% and y = amount at 4.5%.

$$0.05x + 0.045y = 560$$
$$x + y = 12,000$$

Solve equation #2 for x and substitute into equation #1: $x = 12,000 - y$

$$0.05(12,000 - y) + 0.045y = 560$$
$$600 - 0.05y + 0.045y = 560$$
$$-0.005y = -40$$
$$y = 8,000$$

Find x: $x = 12,000 - y = 12,000 - 8,000 = 4,000$

$4,000 is invested at 5% and $8,000 is invested at 4.5%

7. Let x = leotards and y = tights.

$$3x + 4y = 185$$
$$2x + 3y = 127.50$$

Solve by addition: multiply equation #1 by 2 and equation #2 by -3 and add them together.

$$6x + 8y = 370$$
$$\underline{-6x - 9y = -382.50}$$
$$-y = -12.50$$
$$y = 12.50$$

Substitute into equation #1 to find x.

$$3x + 4(12.50) = 185$$
$$3x + 50 = 185$$
$$3x = 135$$
$$x = 45$$

Tights cost $12.50 and a leotard costs $45.00.

9. Let x = pizzas and y = sodas.

Set up two equations: $3x + 4y = 61$ and $2x + 3y = 42$.

Multiply equation #1 by -2; multiply equation #2 by 3, and add them together.

$$-6x - 8y = -122$$
$$\underline{6x + 9y = 126}$$
$$y = 4$$

Substitute to find x: $3x + 4(4) = 61$
$$3x + 16 = 61$$
$$3x = 45$$
$$x = 15$$

A large pizza costs $15 and a pitcher of soda costs $4.

11. Let C = Coca-cola stock and P = Pepsi stock

$$16C + 10P = \$1097.40$$
$$48C + 12P = \$2509.20$$ Multiply equation #1 by -3 and add to equation #2.
$$-48C - 30P = -3292.2$$
$$\underline{48C + 12P = 2509.20}$$
$$-18P = -783$$
$$P = \$43.50$$

Substitute into equation #1: $16C + 10(43.5) = 1097.4$
$$16C + 435 = 1097.4$$
$$16C = 662.4$$
$$C = \$41.40$$

Coke stock is $41.40 per share and Pepsi is $43.50 per share.

13. Let x = 10% acid and y = 25% acid.

Set up two equations: $0.10x + 0.25y = 0.18(30)$

$$x + y = 30$$

Solve equation #2 for y and substitute into equation #1: $y = 30 - x$

$$0.10x + 0.25(30 - x) = 0.18(30)$$
$$0.10x + 7.5 - 0.25x = 5.4$$
$$-0.15x = -2.1$$
$$x = 14$$

She should use 14 mL of 10% and the rest of the 30 mL total is 16 mL of 25%.

15. Let x = 60% silver alloy and y = 40% silver alloy.

Set up two equations: $0.60x + 0.40y = 0.52(20)$ and $x + y = 20$

Solve equation #2 for *y* and substitute into equation #1: $y = 20 - x$

$$0.6x + 0.4(20 - x) = 10.4$$
$$0.6x + 8 - 0.4x = 10.4$$
$$0.2x = 2.4$$
$$x = 12$$

Use 12 g of 60% silver alloy and 8 g of 40% silver alloy.

17. Let x = 10% acid and y = 60% acid.

$$0.10x + 0.60y = 0.40(50)$$
$$x + y = 50$$

Solve equation #2 for x and substitute into equation #1: $x = 50 - y$

$$0.10(50 - y) + 0.60y = 20$$
$$5 - 0.10y + 0.60y = 20$$
$$0.5y = 15$$
$$y = 30$$

Solve for x: $x = 50 - y = 50 - 30 = 20$

Use 20 mL of 10% solution and 30 mL of 60% solution.

19. Let x = peanuts and y = cashews

$$x + y = 26$$
$$2.95x + 9.95y = 132.70$$

Solve equation #1 for x and substitute into equation #2: $x = 26 - y$

$$2.95(26 - y) + 9.95y = 132.7$$
$$76.7 - 2.95y + 9.95y = 132.7$$
$$76.7 + 7y = 132.7$$
$$7y = 56$$
$$y = 8$$

Solve for x: $x = 26 - y = 26 - 8 = 18$

There arc 18 pounds of peanuts and 8 pounds of cashews.

21. Let x = candy @$3.40/lb and y = candy @ $4.60/lb

$$x + y = 10$$
$$3.40x + 4.60y = 4(10)$$

Solve equation #1 for x and substitute into equation #2: $\quad x = 10 - y$

$$3.40(10 - y) + 4.60y = 40$$
$$34 - 3.4y + 4.60y = 40$$
$$34 + 1.2y = 40$$
$$1.2y = 6$$
$$y = 5$$

Solve for x: $x = 10 - y = 10 - 5 = 5$ $\qquad$ He needs 5 lb of each type of candy.

23. The equilibrium price occurs when $D(p) = S(p)$.

$$1200 - 2.8p = 800 + 3.2p$$
$$-6p = -400$$
$$p = 66.67$$

The equilibrium price of the product is $66.67, the price at which supply equals demand.

25. The equilibrium price occurs when $D(p) = S(p)$.

$$9500 - 100p = 5000 + 200p$$
$$-300p = -4500$$
$$p = 15$$

The equilibrium price of the DVD is $15, the price at which supply equals demand.

27. The equilibrium price occurs when $D(p) = S(p)$.

$$30 - 3p = 4p - 5$$
$$35 = 7p$$
$$p = 5$$

The equilibrium price is $5.

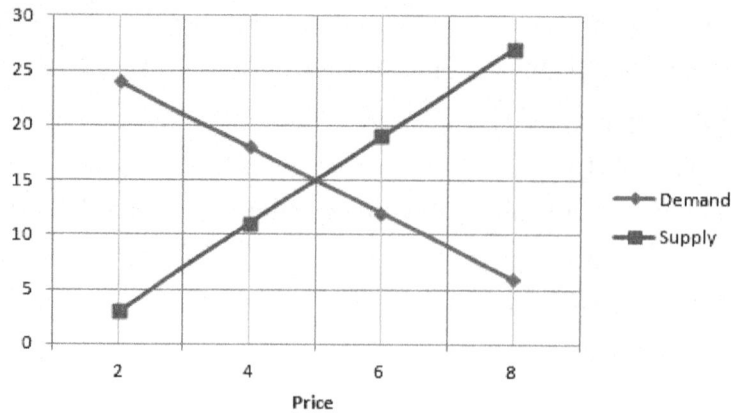

29. Let x = number of backpacks and set sales = costs.

Income from sales: 32x

Production costs: 25,500 + 15x

$$32x = 25,500 + 15x$$
$$17x = 25,500$$
$$x = 1500$$

The company breaks even if they sell 1500 backpacks.

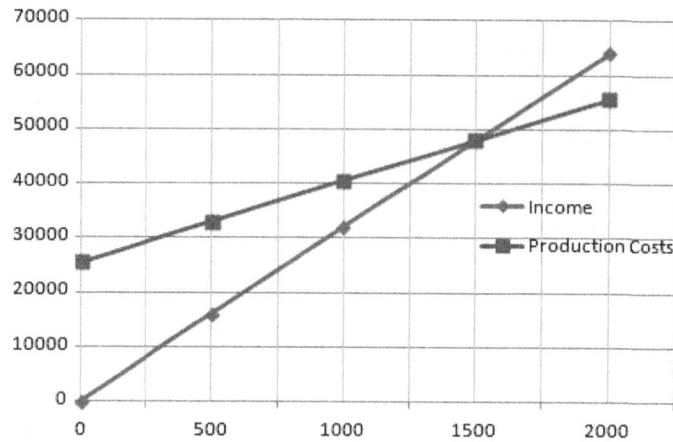

31. Cost at John's: 26 + 0.15x   where x = number of miles driven

Cost at Cookie's: 18 + 0.20x   where x = number of miles driven

$$26 + 0.15x = 18 + 0.20x$$
$$8 = 0.05x$$
$$x = 160$$

You will need to drive 160 miles for the cost of the two agencies to be the same.

To calculate the cost of 200 miles, let x = 200 in each equation.

Cost at John's: 26 + 0.15(200) = 26 + 30 = $56

Cost at Cookie's: 18 + 0.2(200) = 18 + 40 = $58   John's is cheaper.

33. Industrial Paper: 1500 + 0.04x   Paper Factory: 0.12x   where x = monthly sales

$$1500 + 0.04x = 0.12x$$
$$1500 = 0.08x$$
$$x = \$18{,}750 \quad \text{the amount of sales where the salaries are equal}$$

To find the better salary if weekly sales are $5000, let x = 20,000 in each

equation. (x = monthly sales)

Industrial Paper: 1500 + 0.04(20,000) = 1500 + 800 = $2300

Paper Factory :   0.12(20,000) = $2400

He would do better with the $12% straight commission.

35.   Let length = x and width = y.

$$x = 3 + y$$
$$2x + 2y = 14 \qquad \text{Solve by substituting equation \#1 into equation \#2.}$$
$$2(3 + y) + 2y = 14$$
$$6 + 2y + 2y = 14$$
$$6 + 4y = 14$$
$$4y = 8$$
$$y = 2$$

Solve for x: x = 3 + y = 3 + 2 = 5

The length is 5 ft and the width is 2ft.

37. Let length = L and width = W.

$$2L + 2W = 60$$
$$L = W + 6$$

Substitute equation #2 into equation #1.

$$2(W + 6) + 2W = 60$$
$$2W + 12 + 2W = 60$$
$$4W + 12 = 60$$
$$4W = 48$$
$$W = 12$$

Now solve for L in equation #2: L = W + 6 = 12 + 6 = 18

The dimensions are 12 in by 18 in.

39. Let length = L and width = W.

$$2L + 2W = 28$$
$$L = 2W + 2$$ Substitute equation #2 into equation #1.
$$2(2W + 2) + 2W = 28$$
$$4W + 4 + 2W = 28$$
$$6W = 24$$
$$W = 4 \text{ yards}$$

Find L using equation #2

$$L = 2(4) + 2 = 10 \text{ yards}$$ The pen is 4 yards by 10 yards

Practice Set 6-4

1. $y = x^2$ and $3x + y = 10$; substitute equation #1 into equation #2.

Set the equation equal to zero and solve.

$$3x + x^2 = 10$$
$$x^2 + 3x - 10 = 0 \quad \text{factor}$$
$$(x + 5)(x - 2) = 0$$
$$x = -5 \text{ or } 2$$

Substitute both values into equation #1.

If $x = -5$; $y = (-5)^2 = 25$

If $x = 2$; $y = 2^2 = 4$ The solutions are $\{(2, 4), (-5, 25)\}$.

3. $y = x$ and $y = -2x^2 + 1$;

Substitute equation #1 into equation #2, set the equation equal to zero and solve for x.

$$x = -2x^2 + 1$$
$$2x^2 + x - 1 = 0 \quad \text{factor}$$
$$(2x - 1)(x + 1) = 0$$
$$x = -1 \text{ or } 0.5$$

Since $y = x$, the solutions are $\{(-1, -1), (0.5, 0.5)\}$.

5. $y = x^2 - 5x + 6$ and $y = 2x$

Substitute equation #2 into equation #1.

$$2x = x^2 - 5x + 6$$
$$x^2 - 7x + 6 = 0 \quad \text{factor}$$
$$(x - 6)(x - 1) = 0$$
$$x = 6 \text{ or } 1$$

Find y: $y = 2(6) = 12$ and $y = 2(1) = 2$ The solutions are $\{(6, 12), (1, 2)\}$.

136

7. $y = x^2 - 4$ and $y = 2x - 1$

Substitute equation #2 into equation #1.

$$2x - 1 = x^2 - 4$$
$$x^2 - 2x - 3 = 0 \qquad \text{factor}$$
$$(x - 3)(x + 1) = 0$$
$$x = 3 \text{ or } -1$$

Find y: $y = 2(3) - 1 = 5$ and $y = 2(-1) - 1 = -3$ 　　　The solutions are $\{(3, 5), (-1, -3)\}$.

9. $2x + 3y^2 = 2$ and $x - y = 1$

In equation #2: $x = y + 1$

Substitute this expression into equation #1.

$$2(y + 1) + 3y^2 = 2$$
$$3y^2 + 2y + 2 = 2$$
$$3y^2 + 2y = 0$$
$$y(3y + 2) = 0$$
$$y = 0 \text{ or } -\tfrac{2}{3}$$

Find x: $x = 0 + 1 = 0$ and $x = -\tfrac{2}{3} + 1 = \tfrac{1}{3}$ 　　　The solutions are $\{(1,0), (\tfrac{1}{3}, -\tfrac{2}{3})\}$.

11. Add the two equations as they are.

$$x^2 + y^2 = 13$$
$$\underline{x^2 - y^2 = 5}$$
$$2x^2 = 18$$
$$x^2 = 9$$
$$x = 3 \text{ or } -3$$

Solve for y: $3^2 + y^2 = 13$ 　　　　　　Let x = 3.
$$9 + y^2 = 13$$
$$y^2 = 4$$
$$y = 2 \text{ or } -2$$
$$(-3)^2 + y^2 = 13 \qquad \qquad \text{Let x = -3.}$$
$$9 + y^2 = 13$$
$$y^2 = 4$$
$$y = 2 \text{ or } -2$$

The solutions are $\{(3,2)\ (3,-2)\ (-3,2)\ (-3,-2)\}$.

13. $x + y = 7$ and $x^2 + y^2 = 25$; solve for y in equation #1: $y = 7 - x$

Substitute this expression into equation #2.

$$x^2 + (7 - x)^2 = 25$$
$$x^2 + 49 - 14x + x^2 = 25$$
$$2x^2 - 14x + 24 = 0$$
$$x^2 - 7x + 12 = 0 \qquad \text{factor}$$
$$(x - 3)(x - 4) = 0$$
$$x = 3 \text{ or } 4 \qquad\qquad\qquad \text{The numbers are 3 and 4.}$$

15. $x^2 + y^2 = 58$ and $x - y = 10$

Solve equation #2 for x: $x = y + 10$

Substitute this expression into equation #1.

$$(y + 10)^2 + y^2 = 58$$
$$y^2 + 20y + 100 + y^2 = 58$$
$$2y^2 + 20y + 42 = 0$$
$$y^2 + 10y + 21 = 0 \qquad \text{factor}$$
$$(y + 3)(y + 7) = 0$$
$$y = \text{-3 or -7}$$
$$\text{Solve for x: } x = y + 10 = \text{-3} + 10 = 7$$
$$x = y + 10 = \text{-7} + 10 = 3$$

The solutions are 3 and -7 or 7 and -3

17. $xy = 60$ and $y - x = 7$

Solve equation 2 for y: $y = x + 7$

Substitute into equation 1.

$$x(x + 7) = 60$$
$$x^2 + 7x = 60$$
$$x^2 + 7x - 60 = 0 \qquad \text{factor}$$
$$(x - 5)(x + 12) = 0$$
$$x = 5 \text{ or -12}$$

Since the numbers have to be positive, $x = 5$.

Now, solve for y: $y = x + 7 = 5 + 7 = 12$. \qquad The numbers are 12 and 5.

19. $-0.1x^2 - 2x + 90 = 0.1x^2 - x + 30$

Set the equation equal to 0.

$$0.2x^2 + x - 60 = 0$$

Make all coefficients whole numbers by multiplying by 5.

$$x^2 + 5x - 300 = 0 \quad \text{factor}$$
$$(x - 15)(x + 20) = 0$$
$$x = 15 \text{ or } -20$$

The answer must be positive so choose x = 15, or 15,000 pairs of sunglasses.

Now solve for the equilibrium price in either equation.

$$p = 0.1(15)^2 - 15 + 30$$
$$p = 0.1(225) + 15$$
$$p = 22.5 + 15 = \$37.50 \text{ each.}$$

21. $xy = 143$ ft^2 and $2x + 2y = 48$ft

Solve equation #2 for y. $2y = 48 - 2x$

$$y = 24 - x$$

Substitute into equation #1: $x(24 - x) = 143$
$$24x - x^2 = 143$$
$$x^2 - 24x + 143 = 0 \qquad \text{factor}$$
$$(x - 11)(x - 13) = 0$$
$$x = 11 \text{ or } 13$$

The solution is 13 feet by 11 feet.

Chapter 6 Review Problems

1. Substitute into both equations:

   $0 - 2(-2) = 4$ and $2(0) - 2 = -2$ Yes, the pair solves both equations.

2. Substitute into both equations:

   $3(-5) - (-2) = -15 + 2 = -13$

   $-5 - 2(-2) = -5 + 4 = -1$ Yes, the pair solves both equations.

3. Substitute into both equations: $1 + 3 = 4$ and $2(1) - 3 \neq 5$

   No, the pair does not solve both equations.

4. Substitute into both equations: $3(2) + 1 = 7$ and $2 = 2(1)$

   Yes, the pair solves both equations.

Small sketch graphs are very inaccurate in a key like this. Your graph crossing points should closely match the points given here for problem numbers 5-9.

5. (7, 1)

6. no solution; because they don't intersect

7. $(3, 2)$

8. $(1, 3)$

9. $(0, 2)$

10. $3x - y = 7$

   $\underline{2x + y = 3}$ Add the two equations.

   $5x = 10$

   $x = 2$ Substitute this value into equation #2.

   $2(2) + y = 3$

   $4 + y = 3$

   $y = -1$ The solution is $(2, -1)$.

11. $8x - 5y = 32$

   $\underline{4x + 5y = 4}$ Add the two equations.

   $12x = 36$

   $x = 3$ Substitute this value into equation #2.

   $4(3) + 5y = 4$

   $12 + 5y = 4$

   $5y = -8$

   $y = -1.6$ The solution is $(3, -1.6)$.

12. Multiply equation two by 2 and add to equation #1.

   $8x + 2y = 2$

   $\underline{6x - 2y = 12}$

   $14x = 14$

   $x = 1$ Substitute this value into equation #1.

   $8(1) + 2y = 2$

   $2y = -6$

   $y = -3$ The solution is $(1, -3)$.

13. $y = 4x$; Substitute this expression for y in equation.

   $3x + 5(4x) = 69$

   $23x = 69$

   $x = 3$

   Substitute this value into equation #2.

   $y = 4(3) = 12$ The solution is $(3, 12)$.

14. Multiply equation one by -3 and add to equation #2

$-18x - 9y = 0$

$\underline{18x + 9y = 0}$

$0 = 0$ These equations represent the same line.

There are an infinite number of solutions or $\{(x,y) \mid 6x + 3y = 0\}$.

15. Multiply equation #2 by 2 and add to equation #1.

$6x - 8y = 6$

$\underline{-6x + 4y = -4}$

$-4y = 2$

$y = -\dfrac{1}{2}$

Substitute this value into equation #1.

$6x - 8(-\dfrac{1}{2}) = 6$

$6x + 4 = 6$

$6x = 2$

$x = \dfrac{2}{6} = \dfrac{1}{3}$ The solution is $\left(\dfrac{1}{3}, -\dfrac{1}{2}\right)$.

16. Substitute equation #2 into equation #1.

$:2(1 - 5y) + 10y = 3$

$2 - 10y + 10y = 0$

$2 = 0$ (false statement)

These are parallel lines so there is no solution.

17. Multiply equation #2 by -4, rearrange and add to equation #1.

$4x + 5y = 44$

$\underline{-4x + 8y = 8}$

$13y = 52$

$y = 4$

Substitute this value into equation #1.

$4x + 5(4) = 44$

$4x + 20 = 44$

$4x = 24$

$x = 6$ The solution is $(6, 4)$.

18. Multiply equation #1 by 3 and equation #2 by -2 and then add them together.

$6x + 9y = -15$

$\underline{-6x - 8y = 16}$

$y = 1$

Substitute this value into equation #1.

$$2x + 3(1) = -5$$
$$2x = -8$$
$$x = -4$$

The solution is (-4, 1).

19. $x = \dfrac{\begin{vmatrix} 4 & -1 \\ 21 & 1 \end{vmatrix}}{\begin{vmatrix} 2 & -1 \\ 3 & 1 \end{vmatrix}} = \dfrac{4-(-21)}{2-(-3)} = \dfrac{25}{5} = 5$

Substitute this value to solve for y.

$$3(5) + y = 21$$
$$y = 6$$

The solution is (5, 6).

20. $x = \dfrac{\begin{vmatrix} 2 & 2 \\ 7 & 4 \end{vmatrix}}{\begin{vmatrix} 4 & 2 \\ 5 & 4 \end{vmatrix}} = \dfrac{8-14}{16-10} = \dfrac{-6}{6} = -1$

Substitute this value to solve for y.

$$4(-1) + 2y = 2$$
$$2y = 6$$
$$y = 3$$

The solution is (-1, 3).

21. $x = \dfrac{\begin{vmatrix} 2 & 5 \\ 1 & -10 \end{vmatrix}}{\begin{vmatrix} 4 & 5 \\ -8 & -10 \end{vmatrix}} = \dfrac{-20-5}{-40+40} = \dfrac{-25}{0} = \text{undefined}$

Parallel lines with no solution

22. $x = \dfrac{\begin{vmatrix} -4 & 1 \\ 7 & -1 \end{vmatrix}}{\begin{vmatrix} 2 & 1 \\ 1 & -1 \end{vmatrix}} = \dfrac{4-7}{-2-1} = \dfrac{-3}{-3} = 1$

Substitute this value to solve for y.

$$2(1) + y = -4$$
$$y = -6$$

The solution is (1, -6).

142

23. $x = \dfrac{\begin{vmatrix} 0.3 & -0.1 \\ -0.5 & -.02 \end{vmatrix}}{\begin{vmatrix} 0.8 & -0.1 \\ 0.5 & -0.2 \end{vmatrix}} = \dfrac{-0.06 - 0.05}{-0.16 - (-0.05)} = \dfrac{-0.11}{-0.11} = 1$

Substitute this value to solve for y.

$$0.8(1) - 0.1y = 0.3$$
$$-0.1y = -0.5$$
$$y = 5$$

The solution is (1, 5).

24. Algebraically: Solve equation #1 for y and substitute into equation #2: $y = 8 - 2x$

$$5x - 2(8 - 2x) = -16$$
$$5x - 16 + 4x = -16$$
$$9x = 0$$
$$x = 0$$

Solve for y: $y = 8 - 2(0) = 8$ The solution is (0, 8).

By Cramer's rule: $x = \dfrac{\begin{vmatrix} 8 & 1 \\ -16 & -2 \end{vmatrix}}{\begin{vmatrix} 2 & 1 \\ 5 & -2 \end{vmatrix}} = \dfrac{-16 + (16)}{-4 - 5} = \dfrac{0}{-9} = 0$; substitute as above to

solve for y and the solution is (0, 8) as before.

25. Substitute equation two into equation #1 and solve for x by factoring:

$$x^2 + 2x - 3 = 0;$$
$$(x + 3)(x - 1) = 0$$
$$x = -3 \text{ or } 1$$

Substitute both values into equation #2 to get the values of y:

$$y = 2(-3) - 3 = -9$$
$$y = 2(1) - 3 = -1$$

The solutions are $\{(-3, -9), (1, -1)\}$.

26. Substitute equation #2 into equation #1 and solve by factoring.

$$2x^2 + 1 = 3x + 6$$
$$2x^2 - 3x - 5 = 0$$

$$(2x - 5)(x + 1) = 0$$
$$x = -1 \text{ or } 2.5$$

Substitute these values into equation #1 and solve for y.

$$y = 3(-1) + 6 = 3 \text{ and } y = 3(2.5) + 6 = 13.5$$

The solutions are {(2.5, 13.5), (-1, 3)}.

27. Let x = donuts and y = stadium cushions.

$$4x + 6.50y = 318$$
$$x + y = 62$$

Solve the second equation for x and substitute into the first equation. $x = 62 - y$

$$4(62 - y) + 6.50y = 318$$
$$248 - 4y + 6.50y = 318$$
$$2.50y = 70$$
$$y = 28$$

Substitute this value to find x.

$$x = 62 - y = 62 - 28 = 34$$

Tyler sold 34 boxes of donuts and 28 stadium cushions.

28. Let x = 40% solution and y = 70% solution.

$$0.40x + 0.70y = 0.50(120) \text{ and } x + y = 120$$

Solve the second equation for x and substitute into the first equation. $x = 120 - y$

$$0.40(120 - y) + 0.70y = 0.50(120)$$
$$48 - 0.40y + 0.70y = 60$$
$$0.30y = 12$$
$$y = 40$$

Substitute this value to find x

$$x = 120 - y = 120 - 40 = 80$$

We will need 80 mL of the 40% acid solution and 40 mL of the 70% acid solution.

29. Income: 125x

Cost of Production: 20,000 + 25x

$$20,000 + 25x = 125x$$
$$20,000 = 100x$$
$$x = 200$$

200 clocks must be made and sold in order for the company to break even.

144

30. Demand function: $D(p) = 120 - p$

Supply function: $S(p) = 2p - 117$
$$120 - p = 2p - 117$$
$$-3p = -237$$
$$p = 79$$

The equilibrium price is $79.00.

31. Let x = the number of dimes and y = the number of nickels.
$$x + y = 60$$
$$0.10x + 0.05y = 5.75$$

Solve the first equation for x and substitute into the second equation. $x = 60 - y$
$$0.10(60 - y) + 0.05y = 5.75$$
$$6 - 0.10y + 0.05y = 5.75$$
$$-0.05y = -0.25$$
$$y = 5$$

Substitute this value to find x.
$$x = 60 - y = 60 - 5 = 55$$

Cheryl had 55 dimes and 5 nickels.

32. Let x = pounds of potatoes and y = pounds of bananas.
$$10x + 4y = 6.16$$
$$4x + 8y = 6.88$$

Multiply the first equation by -2 and add the resulting equations.
$$-20x - 8y = -12.32$$
$$\underline{4x + 8y = 6.88}$$
$$-16x = -5.44$$
$$x = 0.34$$

Substitute this value to find y.
$$10(0.34) + 4y = 6.16$$
$$3.4 + 4y = 6.16$$
$$4y = 2.76$$
$$y = 0.69$$

Potatoes are $0.34/lb and bananas are $0.69/lb.

33. Income: $5.50x

Cost of Production: $3000 + 3x \qquad$ where x = number of packages of stationery
$$3000 + 3x = 5.50x$$
$$3000 = 2.50x$$
$$x = 1200$$

1200 packages of stationery

34. Let length = L and width = W.

$$2L + 2W = 84$$
$$L = W + 6$$

Substitute equation #2 into equation #1.

$$2(W + 6) + 2W = 84$$
$$2W + 12 + 2W = 84$$
$$4W + 12 = 84$$
$$4W = 72$$
$$W = 18$$

Now solve for L in equation #2: $L = W + 6 = 18 + 6 = 24$

The dimensions are 24ft by 18 ft.

35. answers will vary

Chapter 6 Test

1. $2(-3) + 3(4) = -6 + 12 = 6$

    $2(-3) + 4 = -6 + 4 = -2$

    Therefore, (-3, 4) is the solution to this system of equations.

2. $2(6) - (-1)(-5) = 12 - 5 = 7$

    For numbers 3-5, your graphs should approximate the crossing points given.

3. (4, 1)

4. (2, -1)

5. no solution – parallel lines

6. $y = 4 - 3x$

    $3x + y = 5$ Substitute the first equation into the second and solve for x.

    $3x + (4 - 3x) = 5$

    $4 \neq 5$ no solution – parallel lines

146

7. $x - 2y = 4$

   $3x + 4y = 2$

   Solve the first equation for x and substitute into the second equation: $x = 4 + 2y$

   $3(4 + 2y) + 4y = 2$

   $12 + 6y + 4y = 2$

   $10y = -10$

   $y = -1$

   Substitute this value to find x.

   $x = 4 + 2(-1) = 4 - 2 = 2.$ The solution is (2, -1).

8. $4x + 3y = 15$

   $2x - 5y = 1$

   Multiply the second equation by -2 and add the resulting equations together.

   $4x + 3y = 15$

   $\underline{-4x + 10y = -2}$

   $13y = 13$

   $y = 1$

   Substitute this value for y to find x.

   $4x + 3(1) = 15$

   $4x + 3 = 15$

   $4x = 12$

   $x = 3$ The solution is (3, 1).

9. $2x + 4y = 7$

   $5x - 3y = -2$

   Multiply the first equation by 3 and the second by 4, then add the resulting equations.

   $6x + 12y = 21$

   $\underline{20x - 12y = -8}$

   $26x = 13$

   $x = \dfrac{1}{2}$

   Substitute this value to find y.

   $2(\dfrac{1}{2}) + 4y = 7$

   $1 + 4y = 7$

   $4y = 6$

   $y = \dfrac{3}{2}$ The solution is $\left(\dfrac{1}{2}, \dfrac{3}{2} \right).$

147

10. $x = \dfrac{\begin{vmatrix} 13 & -7 \\ 7 & 5 \end{vmatrix}}{\begin{vmatrix} 3 & -7 \\ 6 & 5 \end{vmatrix}} = \dfrac{65-(-49)}{15-(-42)} = \dfrac{65+49}{15+42} = \dfrac{114}{57} = 2$

Substitute this value to find y.

$$3(2) - 7y = 13$$
$$6 - 7y = 13$$
$$-7y = 7$$
$$y = -1$$

The solution is (2, -1).

11. $x = \dfrac{\begin{vmatrix} 0.3 & -1.2 \\ 1.6 & 1 \end{vmatrix}}{\begin{vmatrix} 0.5 & -1.2 \\ 0.2 & 1 \end{vmatrix}} = \dfrac{0.3-(-1.92)}{0.5-(-0.24)} = \dfrac{0.3+1.92}{0.5+0.24} = \dfrac{2.22}{0.74} = 3$

Substitute this value to find y.

$$0.5(3) - 1.2y = 0.3$$
$$1.5 - 1.2y = 0.3$$
$$-1.2y = -1.2$$
$$y = 1$$

The solution is (3, 1).

12. $2x^2 - y = 0$

$3x - y = -2$

Solve the first equation for y and substitute into the second equation.

$$y = 2x^2$$
$$3x - 2x^2 = -2$$
$$2x^2 - 3x - 2 = 0$$
$$(2x + 1)(x - 2) = 0$$
$$x = -\frac{1}{2} \text{ and } x = 2$$

Substitute to find y.

$$y = 2(-\frac{1}{2})^2 = 2(\frac{1}{4}) = \frac{1}{2} \text{ and } y = 2(2)^2 = 2(4) = 8.$$

$(-\frac{1}{2}, \frac{1}{2})$ and (2, 8) are the solutions to this system of equations.

148

13. Substitute equation 2 into equation 1 and solve by factoring:

$$x + 2(x^2 + 3x + 2) = 4$$
$$2x^2 + 7x = 0$$
$$x(2x + 7) = 0$$
$$x = 0 \text{ or } x = -3.5$$

Substitute both values into equation 2:

$$y = 0^2 + 3(0) + 2 = 2$$
$$y = (-3.5)^2 + 3(-3.5) + 2 = 3.75$$

The solutions are $\{(0, 2) \text{ and } (-3.5, 3.75)\}$.

14. Income: $25x$ Cost of Production: $10,000 + 5x$ where x = number of calculators

$$10,000 + 5x = 25x$$
$$10,000 = 20x$$
$$x = 500$$

500 calculators must be sold for the company to break even.

15. Let x = single rooms and y = double rooms.

$$80x + 100y = 6400$$
$$x + y = 65$$

Solve the second equation for x and substitute into the first equation. $x = 65 - y$

$$80(65 - y) + 100y = 6400$$
$$5200 - 80y + 100y = 6400$$
$$20y = 1200$$
$$y = 60$$

There are 60 double rooms.

16. Let x = 12% alcohol and y = 8% alcohol.

$$0.12x + 0.08y = 0.09(200)$$
$$x + y = 200$$

Solve the second equation for x and substitute into the first equation. $x = 200 - y$

$$0.12(200 - y) + 0.08y = 0.09(200)$$
$$24 - 0.12y + 0.08y = 18$$
$$-0.04y = -6$$
$$y = 150$$

Substitute this value to find x.

$$x = 200 - y = 200 - 150 = 50$$

Use 50 mL of the 12% solution and 150 mL of the 8% solution.

149

17. Let x = amount of bonds at 10.5% and y = amount of bonds at 12%

$$0.105x + 0.12y = 1380$$
$$x + y = 12{,}000$$

Solve the second equation for x and substitute into the first equation. $x = 12{,}000 - y$

$$0.105(12{,}000 - y) + 0.12y = 1380$$
$$1260 - 0.105y + 0.12y = 1380$$
$$0.015y = 120$$
$$y = 8000$$

Substitute this value to find x.

$$x = 12{,}000 - y = 12{,}000 - 8000 = 4000$$

Solution: $4000 at 10.5% and $8000 at 12%

18. Let length = L and width = W.

$$2L + 2W = 76$$
$$L = 2W + 5$$

Substitute equation two into equation one and solve for W.

$$2(2W + 5) + 2W = 76$$
$$4W + 10 + 2W = 76$$
$$6W = 66$$
$$W = 11$$

Substitute this value into equation two.

$$L = 2(11) + 5 = 27$$ The length = 27 in. and the width = 11 in.

19. Supply: $S(p) = 29 + 3p$ Demand: $D(p) = 89 - 7p$ where p = price of the item

$$89 - 7p = 29 + 3p$$
$$60 = 10p$$
$$p = \$6$$ The equilibrium price is $6.

20. Production Costs: $17{,}000 + 6x$

Income: $10x$ where x = number of leotards sold

$$6x + 17000 = 10x$$
$$17000 = 4x$$
$$x = 4250$$

The break-even point occurs when 4250 leotards are sold.

Chapter Seven

Practice Set 7-1

1. True
3. False
5. False
7. False
9. False
11. True
13. A = {x| x is a natural number less than 6}
15. {2, 4, 6, 8, 10, 12, 14,....,40, 42, 44, 46, 48}
17. finite
19. infinite
21. {1, 2, 3, 4, 5, 6, 8}
23. ø or { }

25. {1, 4}

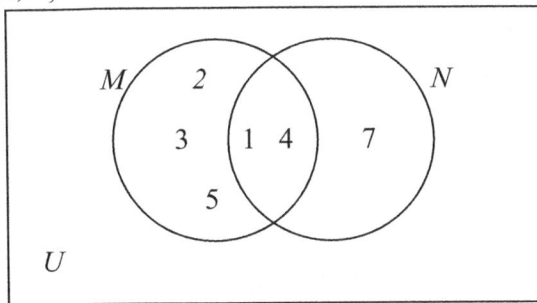

27. {1, 2, 3, 4, 5, 6, 8}

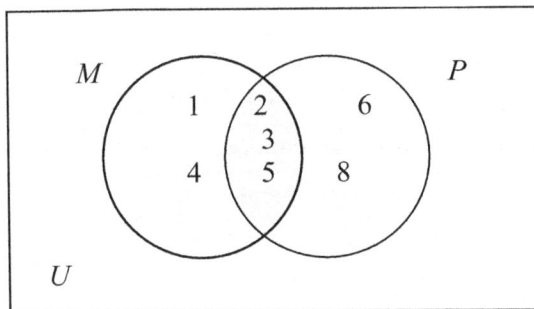

29. {2, 3, 5}

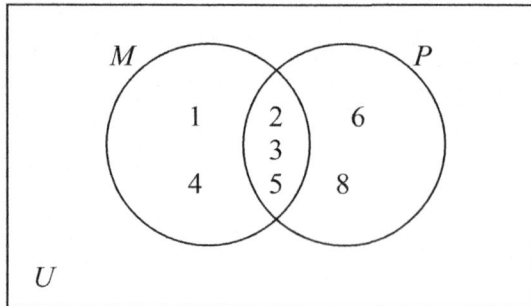

31. {1, 2, 3, 4, 5, 7}

33. {1, 2, 3, 5, 7, 9}

35. {3, 5}

37. Ø or { }

39. A ∩ (B ∩ C) =

    A ∩ {5, 7} =

    {1, 2, 3, 4, 5} ∩ {5, 7} = {5}

Practice Set 7-2

1. An experiment is something you do in order to gain information (data); outcomes are all the possible results of a given experiment; and events are the individual possible outcomes that occur when the experiment is done.

3. The law of large numbers states that the more times an experiment is performed, the more accurately we can predict the probability.

5. independent events; choosing a Democrat does not affect the next choice as a Republican

7. dependent events; choosing a boy as president changes the probability of choosing a boy as vice -president since the number of boys available for vice-president has been reduced by one

9. $\dfrac{155 \text{ Down Syndrome}}{60,000 \text{ births}} = 0.3\%$

11. cardinal: $\dfrac{5}{30} = \dfrac{1}{6} = 16.7\%$; robin: $\dfrac{20}{30} = \dfrac{2}{3} = 66.7\%$

13. $\dfrac{5 \text{ red hair}}{70 \text{ people}} = 7.1\%$

152

15. $\dfrac{41 \text{ made an A}}{672 \text{ students}} = 6.1\%$

17. $\dfrac{8 \text{ sharks}}{72 \text{ fish}} = 11.1\%$

19. $\dfrac{15 \text{ insomnia}}{120 \text{ people}} = 12.5\%$

21. $\dfrac{6 \text{ dizziness}}{120 \text{ people}} = 5\%$; 5% of 25,000 = (0.05)(25,000) = 1250 people

23. $\dfrac{37 \text{ BMI } 20 - 29.9}{55 \text{ members}} = 67.3\%$

25. $\dfrac{9 \text{ BMI} \geq 30.0}{55 \text{ members}} = 16.4\%$

27. 27% (from the table)

29. 4% of African American donors = (0.04)(25) = 1 donor

31. a. $\dfrac{95 \text{ good}}{145 \text{ total}} = \dfrac{19}{29} = 65.5\%$ b. $\dfrac{50 \text{ good}}{65 \text{ lunches}} = \dfrac{10}{13} = 76.9\%$

 c. $\dfrac{35 \text{ poor}}{80 \text{ dinners}} = \dfrac{7}{16} = 43.8\%$ d. $\dfrac{15 \text{ poor}}{65 \text{ lunch}} = \dfrac{3}{13} = 23.1\%$

33. a. $P(ABC \text{ or } NBC) = \dfrac{85}{230} + \dfrac{35}{230} = \dfrac{120}{230} = \dfrac{12}{23} = 52.2\%$

 b. $P(ABC \mid \text{woman}) = \dfrac{50 \text{ women watch ABC}}{95 \text{ women}} = \dfrac{10}{19} = 52.6\%$

 c. $P(ABC \text{ or } NBC \mid \text{man}) = \dfrac{35 \text{ men watch ABC and 25 men watch NBC}}{135 \text{ men}} = \dfrac{60}{135} = \dfrac{4}{9} = 44.4\%$

 d. $P(\text{not CBS} \mid \text{woman}) = \dfrac{75 \text{ women watch ABC or NBC or other}}{95 \text{ women}} = \dfrac{15}{19} = 78.9\%$

Practice Set 7-3

1. Outcomes that have the same chance of occurring in an experiment.

3. No. A theoretical probability of one means that the event is certain to occur.

5. There are 11 letters and 4 s's. $P(s) = \dfrac{4}{11}$

7. There are no a's so $P(a) = 0$.

153

9. $\dfrac{10 \text{ even numbers}}{20 \text{ numbers}} = \dfrac{1}{2}$

11. $\dfrac{1 \text{ three}}{20 \text{ numbers}} = \dfrac{1}{20}$

13. $\dfrac{4 \text{ threes}}{52 \text{ cards}} = \dfrac{1}{13}$

15. $\dfrac{13 \text{ hearts}}{52 \text{ cards}} = \dfrac{1}{4}$

17. 0; There are no spades in the deck that are red.

19. $\dfrac{12 \text{ face cards(K,Q,J)}}{52 \text{ cards}} = \dfrac{3}{13}$

21. $\dfrac{4 \text{ aces}}{52 \text{ cards}} = \dfrac{1}{13}$

23. $\dfrac{2 \text{ cards that are K or Q of hearts}}{52 \text{ cards in a deck}} = \dfrac{1}{26}$

25. $\dfrac{3 \text{ red marbles}}{12 \text{ marbles}} = \dfrac{1}{4}$

27. $\dfrac{7 \text{ marbles that are red or green}}{12 \text{ marbles}} = \dfrac{7}{12}$

29. $\dfrac{1 \text{ five on the die}}{6 \text{ possible outcomes}} = \dfrac{1}{6}$

31. $\dfrac{3 \text{ odd numbers on the die}}{6 \text{ possible outcomes}} = \dfrac{1}{2}$

33. $\dfrac{2 \text{ numbers less than 3}}{6 \text{ numbers on die}} = \dfrac{1}{3}$

35. $\dfrac{2 \text{ successes (either 1 or 4 on the die)}}{6 \text{ possible outcomes}} = \dfrac{1}{3}$

37. $\dfrac{4 \text{ successes (three odd numbers or a six on the die)}}{6 \text{ possible outcomes}} = \dfrac{2}{3}$

39. Since the M&M's are produced with certain percentages of each color, the probability of drawing that particular color will match the composition. Therefore, P(*brown*) = 13% or $\dfrac{13}{100}$.

41. 0, because there are no purple M&M's in this bag.

154

43. $\dfrac{1 \text{ six on the dial}}{10 \text{ possible channels}} = \dfrac{1}{10}$

45. $\dfrac{8 \text{ numbers} < \text{eight on the dial } (0, 1, 2, 3, 4, 5, 6, 7)}{10 \text{ possible channels}} = \dfrac{4}{5}$

47. $\dfrac{1 \text{ correct answer}}{6 \text{ possible answer choices}} = \dfrac{1}{6}$

49. $\dfrac{1 \text{ correct answer}}{4 \text{ possible answer choices}} = \dfrac{1}{4}$

51. $\dfrac{35 \text{ seconds red light}}{90 \text{ seconds for light to cycle through all colors}} = \dfrac{7}{18}$

53. $\dfrac{55 \text{ seconds yellow or green light}}{90 \text{ seconds for light to cycle through all colors}} = \dfrac{11}{18}$

55. $\dfrac{24 \text{ Duracell}}{100 \text{ batteries}} = \dfrac{6}{25} = 0.24$

57. 0, because there are no Kodak batteries in the bin.

Practice Set 7-4

1. odds of a boy = *boy: not a boy* = 1:1

3. odds of winning = *win:not win* = 2:3

5. odds of choosing an ace = *ace:not ace* = 4 : 48 or 1 : 12

7. P(*even*) = $\dfrac{1 \text{ success}}{1 + 1 \text{ outcomes}} = \dfrac{1}{2}$

9. P(*win*) = $\dfrac{3 \text{ wins}}{3 + 11 \text{ outcomes}} = \dfrac{3}{14}$

11. P(*drawing a ten*) = $\dfrac{1 \text{ success}}{1 + 12 \text{ outcomes}} = \dfrac{1}{13}$

13. If the odds of winning are 3:17, that means 3 wins and 17 or losses out of 20 events. The odds against winning are the same as the odds of losing which would be 17:3.

15. If there are 20 members of the class and 3 will be chosen, then 17 will not be chosen. Therefore, the odds of being chosen will be 3:17.

17. a. $\dfrac{1 \text{ five}}{6 \text{ possible outcomes}} = \dfrac{1}{6}$

    b. *1 five:5 not five* results in the odds of rolling a 5 of 1:5

19. Since there are 3 even numbers and 3 odd numbers, the odds against rolling an odd number would be 3 even numbers : 3 odd numbers or 1:1.

155

21. There are 2 numbers less than 3 and 4 numbers that are greater than or equal to 3. Therefore the odds of rolling a number less than 3 are 2:4 or 1:2.

23. a. $\dfrac{13\,\text{clubs}}{52\,\text{cards}} = \dfrac{1}{4}$

   b. Odds of drawing a club would be club : not a club. Therefore, the odds of drawing a club will be 13 *clubs* : 39 *not clubs* or 1:3.

25. Success is drawing one of the four sixes. Failure is drawing a card that is not a six (48 cards). Therefore, the odds of success will be 4 sixes:48 other cards or 1:12.

27. There are 12 face cards in the deck and 40 cards that are not face cards. The odds against drawing a face card are *non-face cards*: *face cards*. Therefore, the odds will be 40: 12 or 10:3.

29. If the P(*overtime*) = $\dfrac{5}{8}$, then on average, out of 8 days, you work overtime 5 times and 3 times you do not. Therefore the odds of working overtime will be 5:3.

31. Out of 1,000,000 raffle tickets, you have 1 ticket and there are 999,999 other tickets. Therefore, the odds of losing will be 999,999 other tickets chosen:1 ticket you bought or 999,999:1.

33. In a horse race, odds of 5:1 give the odds of the horse losing. Therefore, 1 in 6 times the horse wins so P(*win*) = $\dfrac{1}{6}$.

35. If the odds against her being admitted are 9:2, then 9 times out of 11 she will not be admitted and 2 times out of 11 she will be admitted. Therefore, P(*admission*) = $\dfrac{2}{11}$.

37. If the odds against promotion are 5:9, then, on average, 5 times in 14 he will not get a promotion and 9 times in 14 he will get a promotion. Therefore, P(*promotion*) = $\dfrac{9}{14}$.

39. If the odds of winning are 7:5, that means 7 wins : 5 losses. Since there are 12 total events, P(*win*) = $\dfrac{7}{12}$.

41. If the odds of winning are 1:6, that means 1 win : 6 losses. Since there are 7 total events, P(*win*) = $\dfrac{1}{7}$.

43. If the odds against winning are 5:2, that means 5 losses : 2 wins. Since there are 7 total events, P(*win*) = $\dfrac{2}{7}$.

Practice Set 7-5

1. 4 even numbers × 10 numbers × 10 numbers × 10 numbers = 4000 possible four-digit numbers

3. 5 bulbs in the bag × 5 bulbs in the bag × 5 bulbs in the bag = 125 different outcomes possible

5. 5 marbles × 4 marbles = 20 different outcomes possible

7. 2 sides (H,T) × 2 sides × 2 sides = 8 different outcomes possible

156

9. There are 8 possible outcomes with 3 outcomes consisting of one head and two tails: HTT, THT, TTH. Therefore, P(*one head and two tails*) = $\frac{3}{8}$.

11. There are two possible outcomes when tossing a penny {H, T} and six possible outcomes when rolling a die {1, 2, 3, 4, 5, 6}. Therefore, for the two events together there are 2 × 6 = 12 possible outcomes.

13. Only one of the 12 possible outcomes is H3, so P(*H3*) = $\frac{1}{12}$.

15.

1,1	1,2	1,3	1,4	1,5	1,6
2,1	2,2	2,3	2,4	2,5	2,6
3,1	3,2	3,3	3,4	3,5	3,6
4,1	4,2	4,3	4,4	4,5	4,6
5,1	5,2	5,3	5,4	5,5	5,6
6,1	6,2	6,3	6,4	6,5	6,6

a. 36 possible outcomes

b. There are six ways to roll the dice to obtain a total of 7: {1,6; 2,5; 3,4; 4,3; 5,2; 6,1}. Therefore,

P(*sum of 7*) = $\frac{6}{36} = \frac{1}{6}$.

c. 0; There are not two numbers whose sum will give 13.

d. There is only one outcome where 3 is rolled on both dice, so P(*both 3*) = $\frac{1}{36}$.

17. {HH1, HH2, HH3, HH4, HH5, HH6, HT1, HT2, HT3, HT4, HT5, HT6, TH1, TH2, TH3, TH4, TH5, TH6, TT1, TT2, TT3, TT4, TT5, TT6}

19. The outcomes that include HT in any order and a 5 are HT5 and TH5. Therefore,

P(*HT[in any order] and a 5*) = $\frac{2}{24} = \frac{1}{12}$.

21.

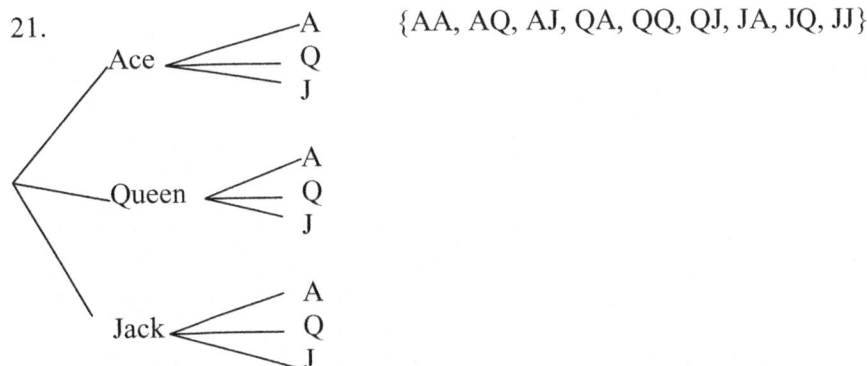

{AA, AQ, AJ, QA, QQ, QJ, JA, JQ, JJ}

23. There is only one outcome out of the 9 possible outcomes with JA. Therefore, $P(JA) = \frac{1}{9}$.

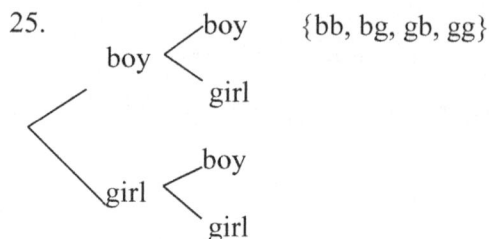

25.

{bb, bg, gb, gg}

27. At least one boy includes all outcomes that have one or more boys: bg, gb, bb.

Therefore, $P(\textit{at least one boy}) = \frac{3}{4}$.

29. There are 26 letters in the alphabet and 10 numbers available (0-9). To calculate the number of possible passwords, multiply the number of outcomes for each event together.

Letter, letter, letter, number, number gives $26 \times 26 \times 26 \times 10 \times 10 = 1{,}757{,}600$ possible passwords.

31. If there is only one possible successful password out of the number of possible passwords (see problem 29) then $P(\textit{guessing correct password on the first guess}) = \frac{1}{1{,}757{,}600}$.

33. Possible combinations for three letters include 3 choices for first letter × 2 choices for second letter × 1 choice for third letter = 6 possible "words" using 3 letters.

Practice Set 7-6

1. $P(K \text{ or } Q) = \frac{4}{52} + \frac{4}{52} = \frac{8}{52} = \frac{2}{13}$; Yes, it is impossible to pick one card and get a king and a queen at the same time.

3. $P(\textit{black or spade}) = \frac{26}{52} + \frac{13}{52} - \frac{13}{52} = \frac{26}{52} = \frac{1}{2}$; No, it is possible to pick one card and get a black spade because all spades are black!

5. $P(\textit{face card or black card}) = \frac{12}{52} + \frac{26}{52} - \frac{6}{52} = \frac{32}{52} = \frac{8}{13}$; No; it is possible to pick one card and get a black face card. (jack of spades and clubs, queen of spades and clubs, king of spades and clubs).

7. $P(2 \text{ or a number} > 2) = \frac{1}{8} + \frac{6}{8} = \frac{7}{8}$

9. $P(\textit{odd or a number} > 2) = \frac{3}{6} + \frac{4}{6} - \frac{2}{6} = \frac{5}{6}$

11. $P(\textit{odd number or even number}) = 1$; all numbers on a die are either even or odd

158

13. *P(rolling 1 or 6)* = ⅓; 2 successes out of 6 possible outcomes gives 2 out of 6 or ⅓ probability

15. $P(male) = \dfrac{110 \text{ males}}{205 \text{ students}} = \dfrac{22}{41}$

17. $P(automotive \text{ or } early childhood) = \dfrac{35 \text{ automotive students}}{205 \text{ students}} + \dfrac{50 \text{ early childhood students}}{205 \text{ students}} = \dfrac{86}{205} = \dfrac{17}{41}$

19. *P(male or criminal justice)* =

$$\dfrac{110 \text{ male students}}{205 \text{ students}} + \dfrac{73 \text{ criminal justice students}}{205 \text{ students}} - \dfrac{48 \text{ male criminal justice students}}{205 \text{ students}} = \dfrac{135}{205} = \dfrac{27}{41}$$

21. $P(approves \text{ } death penalty) = \dfrac{715 \text{ approve}}{1140 \text{ polled}} = \dfrac{143}{228} = 62.7\%$

23. $P(from \text{ } East \text{ } or \text{ } South) = \dfrac{299 \text{ from East}}{1140 \text{ polled}} + \dfrac{250 \text{ from South}}{1140 \text{ polled}} = \dfrac{549}{1140} = \dfrac{183}{380} = 48.2\%$

25. *P(from South or disapproves of death penalty)* =

$$\dfrac{250 \text{ from South}}{1140 \text{ polled}} + \dfrac{375 \text{ disapprove}}{1140 \text{ polled}} - \dfrac{70 \text{ from South disapprove}}{1140 \text{ polled}} = \dfrac{555}{1140} = \dfrac{37}{76} = 48.7\%$$

27. $P(approves \text{ } of \text{ } death penalty \mid from \text{ } South) = \dfrac{165 \text{ from South approve}}{250 \text{ from South}} = \dfrac{33}{50} = 66\%$

29. $P(freshman \text{ } or \text{ } female) = \dfrac{25 \text{ freshmen}}{60 \text{ students}} + \dfrac{27 \text{ females}}{60 \text{ students}} - \dfrac{15 \text{ female freshmen}}{60 \text{ students}} = \dfrac{37}{60} = 61.7\%$

Practice Set 7-7

1. Independent; Because the card is returned, the probability of the second event occurring after the first event has occurred does not change.

3. Dependent; Because the first marble is not returned to the bag, the probability for the second draw is changed since the total number of possible outcomes is reduced.

5. Independent; The outcomes from rolling a die and flipping a coin have no bearing on each other or successive events.

7. $\dfrac{42 \text{ kings}}{52 \text{ cards}} \times \dfrac{4 \text{ kings}}{52 \text{ cards}} = \dfrac{1}{13} \times \dfrac{1}{13} = \dfrac{1}{169}$

159

9. $\dfrac{4 \text{ red marbles}}{9 \text{ total marbles}} \times \dfrac{5 \text{ blue marbles}}{8 \text{ total marbles}} = \dfrac{20}{72} = \dfrac{5}{18}$

11. $\dfrac{1 \text{ four on a die}}{6 \text{ possible outcomes}} \times \dfrac{1 \text{ head on a coin}}{2 \text{ possible outcomes}} = \dfrac{1}{12}$

13. Because the flowers are planted, the selections are done without replacement,

$P(white \text{ and } white \text{ and } white) = \dfrac{12}{24} \times \dfrac{11}{23} \times \dfrac{10}{22} = \dfrac{5}{46} = 10.9\%$

15. Because the flowers are planted, the selections are done without replacement,

$P(not\ red \text{ and } not\ red \text{ and } not\ red) = \dfrac{17}{24} \times \dfrac{16}{23} \times \dfrac{15}{22} = \dfrac{85}{253} = 33.6\%$

17. $P(lose \text{ and } win \text{ and } win \text{ and } lose \text{ and } lose) = \dfrac{3}{5} \times \dfrac{2}{5} \times \dfrac{2}{5} \times \dfrac{3}{5} \times \dfrac{3}{5} = \dfrac{108}{3125} = 3.5\%$

19. $P(win \text{ and } win \text{ and } win \text{ and } win \text{ and } win) = \dfrac{2}{5} \times \dfrac{2}{5} \times \dfrac{2}{5} \times \dfrac{2}{5} \times \dfrac{2}{5} = \dfrac{32}{3125} = 1.0\%$

21. There are 7 consonants and 4 vowels in the word *Mississippi*. Selection is done without replacement.

   a. $P(consonant \text{ and } consonant \text{ and } consonant) = \dfrac{7}{11} \times \dfrac{6}{10} \times \dfrac{5}{9} = \dfrac{21}{99} = 0.212121... = 21.2\%$

   b. $P(vowel \text{ and } consonant \text{ and } vowel) = \dfrac{4}{11} \times \dfrac{7}{10} \times \dfrac{3}{9} = \dfrac{14}{165} = 0.084848... = 8.5\%$

23. For a single birth, $P(boy) = \dfrac{1}{2}$. For three births, $P(three\ boys) = \dfrac{1}{2} \times \dfrac{1}{2} \times \dfrac{1}{2} = \dfrac{1}{8} = 0.125$

25. For a single birth, $P(boy) = \dfrac{1}{2}$ and $P(girl) = \dfrac{1}{2}$.

   For three births, $P(a\ boy,\ then\ a\ girl,\ then\ a\ girl) = \dfrac{1}{2} \times \dfrac{1}{2} \times \dfrac{1}{2} = \dfrac{1}{8} = 0.125$

27. $\dfrac{12 \text{ males}}{27 \text{ members}} \times \dfrac{11 \text{ remaining males}}{26 \text{ remaining members}} \times \dfrac{10 \text{ remaining males}}{25 \text{ remaining members}} = 7.5\%$

29. $\dfrac{1}{365} \times \dfrac{1}{365} = \dfrac{1}{133,225}$

Practice Set 7-8

1. $_7C_2 = \dfrac{7!}{2!5!} = \dfrac{7 \cdot 6 \cdot 5 \cdot 4 \cdot 3 \cdot 2 \cdot 1}{(2 \cdot 1)(5 \cdot 4 \cdot 3 \cdot 2 \cdot 1)} = 21$

3. $_7P_2 = \dfrac{7!}{5!} = \dfrac{7 \cdot 6 \cdot 5 \cdot 4 \cdot 3 \cdot 2 \cdot 1}{(5 \cdot 4 \cdot 3 \cdot 2 \cdot 1)} = 42$

5. $9! = 9 \cdot 8 \cdot 7 \cdot 6 \cdot 5 \cdot 4 \cdot 3 \cdot 2 \cdot 1 = 362,880$

7. Only positive whole numbers have a factorial value. Thus, there is no value here.

9. In a combination, order is unimportant, so that 1,2 and 2,1 are the same combination of the numbers 1 and 2. In a permutation, order is important, so that 1,2 and 2,1 are different permutations (orderings) of the numbers 1 and 2.

11. If Jane must be on the team, then the other three members will be chosen from the nine remaining girls, and order doesn't matter here. $_9C_3 = 84$.

13. Because order is important when batting, this is a permutation. You can calculate it using the counting principle or a permutation. $9! = 362,880$ or $_9P_9 = 362,880$.

15. A person may choose no toppings, or 1 or 2 or...etc. So, calculate all possibilities and add.

    no toppings + 1 topping + 2 toppings + 3 toppings + 4 toppings + 5 toppings

    $1 + {}_5C_1 + {}_5C_2 + {}_5C_3 + {}_5C_4 + {}_5C_5 = x$

    $1 + 5 + 10 + 10 + 5 + 1 = 32$

17. Order is important in this problem, so a permutation can be used. You can also use the counting principle. $4! = 24$ or $_4P_4 = 24$.

19. a. Because order is important in a word, this is a permutation. You can calculate it using the counting principle or a permutation. $9! = 362,880$ or $_9P_9 = 362,880$

    b. Choosing 4 letters out of the nine to create different words by changing their order results in $_9P_4 = 3024$ different words.

21. a. Because order is important in a ship's message, this is a permutation. You can calculate it using the counting principle or a permutation. $8! = 40,320$ or $_8P_8 = 40,320$.

    b. Because order is important in a ship's message, this is a permutation. You can calculate it using the counting principle or a permutation. $8 \cdot 7 \cdot 6 = 336$ or $_8P_3 = 336$.

    c. Because order is important in a ship's message, this is a permutation. You can calculate it using the counting principle or a permutation. $8 \cdot 7 \cdot 6 \cdot 5 = 1680$ or $_8P_4 = 1680$.

23. Use the counting principle to calculate the total number of possible outcomes.

    8 entries $\cdot$ 9 entries = 72 daily-double tickets to purchase to guarantee a win.

25. Since any one of 10 numbers can be repeated in a social security number having 9 digits, there are $10^9 = 1,000,000,000$ possible social security numbers. (Remember that a permutation cannot be used if repetition of digits is allowed, so the counting principle must be used here.)

27. Since order is not important in choosing cars, a combination will give the answer. $_{10}C_6 = 210$.

29. Since order is not important in choosing finalists, a combination will give the answer. $_9C_3 = 84$.

31. You are choosing 3 marbles out of 9 and you want to find the probability that all are red.

    $$P(3\ red) = \frac{\text{choosing 3 red out of 5}}{\text{choosing 3 out of 9}} = \frac{{}_5C_3}{{}_9C_3} = \frac{10}{84} = \frac{5}{42} \text{ or } P(3\ red) = \frac{5}{9} \times \frac{4}{8} \times \frac{3}{7} = \frac{5}{42}.$$

161

33. You are choosing 2 numbers out of 6 and you want to find the probability that both are even.

$$P(2\ even) = \frac{\text{choosing 2 even out of 3}}{\text{choosing 2 out of 6}} = \frac{_3C_2}{_6C_2} = \frac{3}{15} = \frac{1}{5} \text{ or } P(2\ even) = \frac{3}{6} \times \frac{2}{5} = \frac{1}{5}.$$

35. You are choosing 3 batteries out of 8 and you want to find the probability that all are good.

$$P(3\ good) = \frac{\text{choosing 3 good out of 4}}{\text{choosing 3 out of 8}} = \frac{_4C_3}{_8C_3} = \frac{4}{56} = \frac{1}{14} \text{ or } P(3\ good) = \frac{4}{8} \times \frac{3}{7} \times \frac{2}{6} = \frac{1}{14}.$$

37. You are choosing 2 bills out of 8 and you want to find the probability that both are \$5 bills.

$$P(two\ \$5\ bills) = \frac{\text{choosing two \$5 bills out of 4}}{\text{choosing 2 out of 8}} = \frac{_4C_2}{_8C_2} = \frac{6}{28} = \frac{3}{14} \text{ or } P(two\ \$5\ bills) = \frac{4}{8} \times \frac{3}{7} = \frac{3}{14}.$$

39. a. $_6P_6 = 720$ or using the counting principle: $6 \cdot 5 \cdot 4 \cdot 3 \cdot 2 \cdot 1 = 720$

b. $_6P_4 = 360$ or using the counting principle: $6 \cdot 5 \cdot 4 \cdot 3 = 360$

41. $$P(aisle\ seats) = \frac{\text{get 4 aisle seats out of 6}}{\text{get 4 seats out of 10 empty seats}} = \frac{_6C_4}{_{10}C_4} = \frac{15}{210} = \frac{1}{14} \text{ or}$$

$$P(aisle\ seats) = \frac{6}{10} \times \frac{5}{9} \times \frac{4}{8} \times \frac{3}{7} = \frac{1}{14}.$$

43. a. The probability that the lot will pass is based on the probability of choosing 3 good chips from the

98 that are not defective. $P(3\ good\ chips) = \frac{\text{choosing 3 good chips out of 98}}{\text{choosing 3 out of 100}} = \frac{_{98}C_3}{_{100}C_3} = 0.940606$

$= 94.1\%$ or $P(3\ good\ chips) = \frac{98}{100} \times \frac{97}{99} \times \frac{96}{98} = 0.940606 = 94.1\%$

b. answers will vary

45. $$P(uncle\ arrives\ first) = \frac{2\ \text{uncles}}{10\ \text{family members}} = \frac{1}{5}$$

47. $$P(four\ instructors) = \frac{\text{choose 4 out of 8 instructors}}{\text{choose 4 out of 15}} = \frac{_8C_4}{_{15}C_4} = \frac{70}{1365} = \frac{2}{39} = 0.05128$$

49. $$P(3\ numbers > 4) = \frac{\text{choosing 3 numbers} > 4 \text{ out of 5 possible}}{\text{choosing 3 out of 10}} = \frac{_5C_3}{_{10}C_3} = \frac{10}{120} = \frac{1}{12} \text{ or}$$

$$P(3\ numbers > 4) = \frac{5}{10} \times \frac{4}{9} \times \frac{3}{8} = \frac{1}{12}.$$

51. There are 4 digits beginning with a five followed by three digits in each extension. Therefore, there are 10 choices for each of the three random digits so, using the counting principle, the total number of possible extensions is $10^3 = 1000$.

53. various answers

Chapter 7 Review Problems

1.

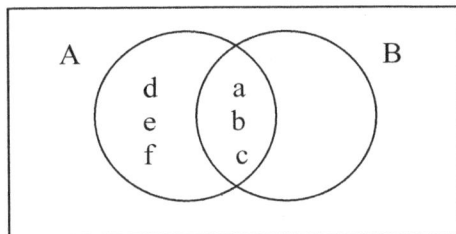

2. {a, b, c, e, g}

3. {f}

4. { } or ø

5. {a, c, e, f, g, h}

6. No. All elements in A are not contained in B.

7. No. All elements in C are not contained in A.

8. Yes; there are six elements in set A so it is finite.

9. No; there are two elements in set D so it is finite.

10. A probability is a mathematical calculation of the likelihood of a given event occurring in preference to all other possible events that could occur in a given situation (or experiment). In other words, it is the proportion (or fraction) of times that a particular outcome will occur.

11. In problems where several outcomes are possible, use the addition rule if the word "or" is used to connect the events and use the multiplication rule if the word "and" is used.

12. A probability is the proportion of outcomes considered favorable compared to the total number of possible results. The odds of an event compares outcomes considered favorable to the number of outcomes considered unfavorable.

13. There are 5 combinations that will total 8: {2,6} {3,5} {4,4} {5,3} {6,2}. So *P(sum of 8)* = $\frac{5}{36}$.

14. 0 (There are no combinations that will total 13.)

15. There are 15 combinations that give sums less than 7: {1,5} {1,4} {1,3} {1,2} {1,1} {2,1} {2,2}

   {2,3} {2,4} {3,1} {3,2} {3,3} {4,1} {4,2} {5,1}. *P(sum < 7)*= $\frac{15}{36} = \frac{5}{12}$.

16. *P(red)* = $\frac{7}{20}$

17. *P(green or blue)* = $\frac{8}{20} + \frac{5}{20} = \frac{13}{20}$

18. *P(three blue)* = $\frac{5}{20} \times \frac{4}{19} \times \frac{3}{18} = \frac{1}{114}$

19. *P(male who drinks root beer)* = $\frac{63 \text{ male drinkers}}{200 \text{ students}} = \frac{63}{200} = 31.5\%$

163

20. $P(\text{female nondrinker}) = \dfrac{48 \text{ females nondrinker}}{200 \text{ students}} = \dfrac{6}{25} = 24\%$

21. $P(\text{drinker} \mid \text{male student}) = \dfrac{63 \text{ male drinkers}}{113 \text{ males}} = \dfrac{63}{113} = 55.8\%$

22. $P(\text{woman} \mid \text{nondrinker}) = \dfrac{48 \text{ women nondrinker}}{98 \text{ nondrinkers}} = \dfrac{24}{49} = 49.0\%$

23. There are three face cards (K, Q, J) per suit so there are 12 face cards in the deck. This means that there are 40 non-face cards in the deck. The odds in favor of drawing a face card are *face card:non-face card* = 12:40 = 3:10.

24. $10^9 = 1$ billion

25. Yes, because by 2050 the population will be well over one billion.

26. $P(\text{good part}) = \dfrac{73}{75}$ or about a 97% chance of choosing a good one out.

27. $P(\text{rainy day}) = \dfrac{92}{365}$ or about a 25% chance of a rainy day.

28. $P(\text{ABC or NBC}) = \dfrac{80}{215} + \dfrac{30}{215} = \dfrac{110}{215} = \dfrac{22}{43} = 51.2\%$

29. $P(\text{ABC or a woman}) = \dfrac{80 \text{ watch ABC}}{215} + \dfrac{100 \text{ women}}{215} - \dfrac{50 \text{ women watch ABC}}{215} = \dfrac{130}{215} = \dfrac{26}{43} = 60.5\%$

30. $P(\text{watches Fox} \mid \text{man}) = \dfrac{25 \text{ men watch Fox}}{115 \text{ men}} = \dfrac{5}{23} = 21.7\%$

31. $_7P_3 = 210$ or use the counting principle: $7 \times 6 \times 5 = 210$.

32. Two teachers and three students gives $_6C_2 \cdot {}_{50}C_3 = 294,000$ possible committees.

33. Since order is not important in choosing flavors, a combination will give the answer. $_{15}C_3 = 455$.

34. He has 8 possible successful tickets and there are 1492 other tickets. His odds of winning are 8:1492 or 2:373.

35. $4! = 4 \times 3 \times 2 \times 1 = 24$

36. Factorials do not apply to negative numbers. The answer is undefined.

37. Factorials do not apply to decimals. The answer is undefined.

38. $_7C_3 = \dfrac{7!}{3!\,4!} = \dfrac{7 \cdot 6 \cdot 5 \cdot 4 \cdot 3 \cdot 2 \cdot 1}{(3 \cdot 2 \cdot 1)(4 \cdot 3 \cdot 2 \cdot 1)} = 35$

39. $_7P_5 = \dfrac{7!}{2!} = \dfrac{7 \cdot 6 \cdot 5 \cdot 4 \cdot 3 \cdot 2 \cdot 1}{(2 \cdot 1)} = 2520$

40. $_{23}C_3 = \dfrac{23!}{3!\,20!} = \dfrac{23 \cdot 22 \cdot 21 \cdot 20 \cdots 3 \cdot 2 \cdot 1}{(3 \cdot 2 \cdot 1)(20 \cdot 19 \cdot 18 \cdots 2 \cdot 1)} = 1771$

Chapter 7 Test

1.

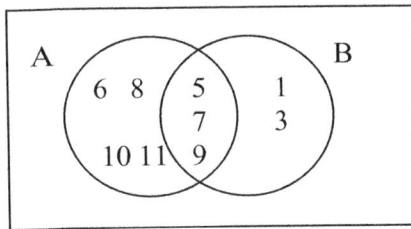

2. $\{5, 7, 9\}$

3. $\{ \ \}$ or ø

4. $\{10, 11, 12, 13, 15\}$

5. S = {H1, H2, H3, H4, H5, H6, T1, T2, T3, T4, T5, T6}

6. $P(head \ and \ 5 \ on \ die) = \dfrac{1}{2} \times \dfrac{1}{6} = \dfrac{1}{12}$

7. $P(tail \ and \ even) = \dfrac{1}{2} \times \dfrac{3}{6} = \dfrac{1}{4}$

8. $P(adequate \ fire \ protection) = \dfrac{85 \ \text{adequate protection}}{150 \ \text{residents}} = \dfrac{17}{30} = 56.7\%$

9. $P(adequate \mid county \ resident) = \dfrac{29 \ \text{county who believe adequate}}{60 \ \text{county residents}} = \dfrac{29}{60} = 48.3\%$

10. $P(inadequate \mid city \ resident) = \dfrac{34 \ \text{city who believe inadequate}}{90 \ \text{city residents}} = \dfrac{17}{45} = 37.8\%$

11. Using the counting principle, three numbers and three letters with no replacement gives

$8 \times 9 \times 8 \times 26 \times 25 \times 24 = 8{,}985{,}600$ possible codes.

12. There are 3 even numbers on the die and 3 odd numbers, so the odds are *even:odd* or 3:3 = 1:1.

13. $P(even \ number) = \dfrac{3 \ \text{even numbers}}{6 \ \text{possible outcomes}} = \dfrac{1}{2}$

14. *P(Doris is promoted)* = $1 - 0.25 - 0.34 = 0.41$

15. Using the counting principle, $3 \times 2 \times 1 = 6$

16. 1 (oat)

17. $P(meaningful \ word) = \dfrac{1}{6}$

18. Only nine plates out of all possible plates will have all four numbers the same.

There are $9 \times 9 \times 9 \times 9 = 6561$ different possible plates. $P(4 \ numbers \ alike) = \dfrac{9}{6561} = \dfrac{1}{729}$

19. $P(red) = \dfrac{26 \text{ red cards}}{52 \text{ cards}} = \dfrac{1}{2}$

20. $P(queen) = \dfrac{4 \text{ queens}}{52 \text{ cards}} = \dfrac{1}{13}$

21. $P(drawing\ 3\ of\ diamonds\ twice\,) = \dfrac{1}{52} \times \dfrac{1}{52} = \dfrac{1}{2704}$

22. $P(face\ card\ or\ diamond) = \dfrac{12}{52} + \dfrac{13}{52} - \dfrac{3}{52} = \dfrac{22}{52} = \dfrac{11}{26}$

23. $P(Ringo\ wins) = \dfrac{1}{4}$ $P(Bonnie\ wins) = \dfrac{1}{16}$

    $P(Clyde\ wins) = \dfrac{5}{13}$ $P(Zero\ wins) = \dfrac{2}{5}$

24. There are only two choices for children, boy or girl, each time, so for four births, $2 \times 2 \times 2 \times 2 = 16$ possible combination of boys and girls. But because order is not specified, many of the groupings are the same. If you list all possible combinations, 6 of the 16 will contain two boys and two girls.

    $P(family\ with\ two\ boys\ and\ two\ girls) = \dfrac{6}{16} = \dfrac{3}{8}.$

25. Order is important in this problem, so a permutation can be used. You can also use the counting principle. $6! = 720$ or $_6P_6 = 720$.

26. If an event is certain, the probability that it won't occur is 0. In other words, there is no chance that it *won't* occur because it is certain to occur.

27. $P(missing\ a\ question) = \dfrac{1}{2}$; $P(missing\ 10\ questions) = \left(\dfrac{1}{2}\right)^{10} = \dfrac{1}{1024}$

28. $P(no\ improvement) = \dfrac{24 \text{ no improvement}}{60 \text{ couples}} = \dfrac{2}{5} = 40\%$

29. $P(couple\ with\ children\,/\,couple\ reported\ improvement\,) = \dfrac{27}{36} = \dfrac{3}{4} = 75\%$

30. $P(couple\ reported\ improvement\ \&\ has\ no\ children) = \dfrac{9 \text{ reported improvement \& have no children}}{60 \text{ couples}} =$

    $\dfrac{3}{20} = 15\%$

Chapter Eight

Practice Set 8.1

1. Inferential; The results are based on a survey.

3. Descriptive; Though it may not be exact, it is still descriptive of the grades over a three year period.

5. Descriptive; Exact calculations are done after every major league game using the population of data.

7. Descriptive: This data is easily available in college records and is based on a complete population of data.

9. Descriptive; This number is based on gate counts for all attendees at each game.

11. numerical data; quantitative

13. numerical data; quantitative

15. non-numerical data; qualitative

17. Numbers are used but they represent personal preferences which is non-numerical data; qualitative

19. numerical data; quantitative

21. High schools were chosen based on geographical location, so this is a cluster sample.

23. Each student has an equal likelihood of being chosen, so it is simple random sample.

25. This survey is easy to do with no real plan or sequence and students are readily available, so this is a convenience sample.

27. Since two classes are chosen from all classes being taught by the instructor and all students in those classes are surveyed, this is a cluster sample.

29. All members have an equal chance of being chosen, so this is a simple random sample.

31. The numbering scale on the vertical axis exaggerates the differences.

33. Ask how many dentists were polled.

35. The number seems too precise for a general statement.

37. There is a misuse of a percent with a number greater than 100%.

39. The survey is not representative of the general population, just senior citizens.

Practice Set 8.2

1. a.

Classes	Frequency
50-54	1
45-49	2
40-44	3
35-39	4
30-34	6
25-29	20
20-24	12
15-19	2

b.

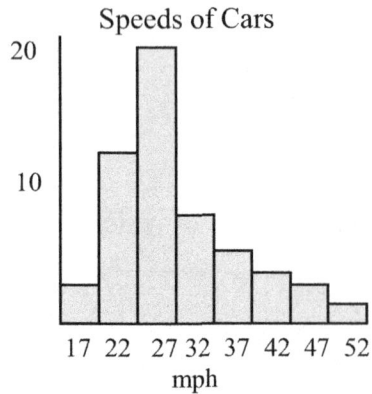

Speeds of Cars

c. This graph is skewed right with more data with low values and relatively few high speeds.

3. a.

Classes	Frequency
500-549	1
450-499	0
400-449	3
350-399	3
300-349	4
250-299	2
200-249	8
150-199	7
100-149	10
50-99	10
0-49	6

b.

Weights of Bears

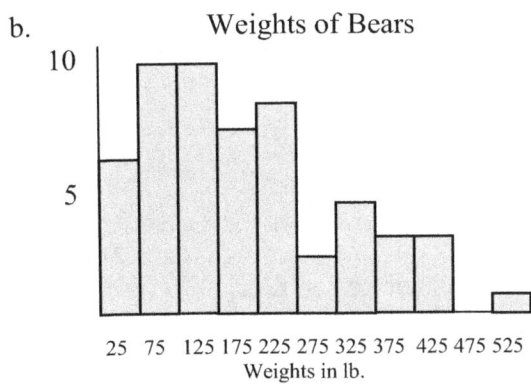

Weights in lb.

5. a.

Classes	Frequency
60-66	3
53-59	7
46-52	3
39-45	4
32-38	3
25-31	9
18-24	7

b.

Age Distribution-History Class

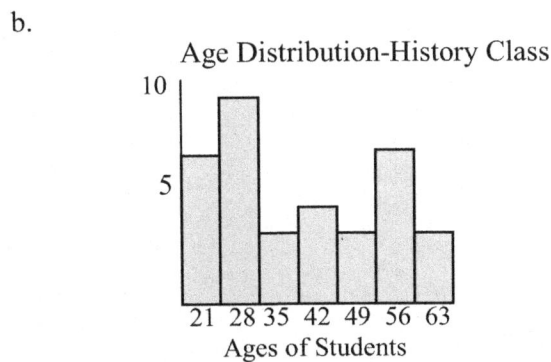

Ages of Students

169

7a.

Classes	Frequency
52-57	1
46-51	1
40-45	4
34-39	5
28-33	14
22-27	21
16-21	4

b.

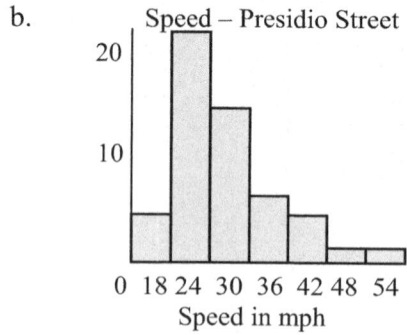

Speed – Presidio Street

c.

```
5│ 2
4│ 13538
3│ 4162087253
2│ 79133282328174993578476658958536
1│ 86
```

9a.

Classes	Frequency
6-11	2
12-18	3
19-25	7
26-32	16
33-39	3
40-46	6
47-53	2
54-60	1

b. see answer key in text

c. normal

Practice Set 8.3

1.

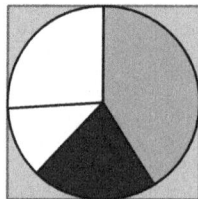

kitchen
bathroom
bedroom
other

170

Kitchen: $\dfrac{369}{900} = 0.41 = 41\%$ of the circle

  Angle: 41% of 360° = 0.41 × 360 = 148°

Bathroom: $\dfrac{189}{900} = 0.21 = 21\%$ of the circle

  Angle: 21% of 360° = 0.21 × 360 = 76°

Bedroom: $\dfrac{108}{900} = 0.12 = 12\%$ of the circle

  Angle: 12% of 360° = 0.12 × 360 = 43°

Other: $\dfrac{234}{900} = 0.26 = 26\%$ of the circle

  Angle: 26% of 360° = 0.26 × 360 = 94°

3. a.
0	8
1	2 2 2 2 5 5 6 6 7 7 8 8 8 9 9
2	0 1 2 3 3 6
3	0 5
4	4

   b. Most students took between 12 and 19 minutes to complete the survey.

5. a.   horizontal axis – months
        vertical axis – rainfall amounts

   b.   October – March

   c.   February – less rainfall

7. a.
1	8 8 9 9
2	0 0 2 2 2 3 3 5 7
3	
4	2 3
5	7

b. The majority of students are in their twenties.

9. a. 3% are from Africa.

b. Yes, because $36\% > 33.3\%$ or $\dfrac{1}{3}$

c. 49% of the total = (0.49)(7,500,000) = 3,675,000

d. Oceania has the lowest percentage based on the graph.

Practice Set 8.4

1. Various answers will be given but should reflect an understanding of the meaning of "average".

3. sum = 46, n = 10, numbers sorted for finding the median and mode: 2, 2, 2, 3, 3, 4, 6, 7, 8, 9; mean = 4.6, median = 3.5, mode = 2; The best average is the mean since the mean and median are close to the same value.

5. sum = 251, n = 20, sorted list: 2,2,3,3,3,4,4,5,5,6,6,7,8,8,9,9,9,9,72,77 mean = 12.6, median = 6, mode = 9, The best average is the median since the mean and the median vary greatly due to the two very large values in the list.

7. sum = 43.79, n = 12, sorted list: 0.35, 1.23, 1.54, 2.65, 2.69, 2.85, 2.88, 3.21, 3.27, 4.25, 6.88, 11.99; mean = 3.65, median = 2.865, no mode; The best average is the median since the mean and the median vary greatly due to the one large value in the list.

9. A few persons in the US, like Bill Gates, Oprah Winfrey, and others make so much money in a year that they warp the mean average income for the entire nation.

11. a. the mode since you would like to see where "most" family sizes fall
    b. the mean because you want to be sure to include everyone who is a student
    c. the median was probably used to avoid having the average price warped by a few multimillion dollar homes
    d. the mean because the national debt would be divided by the total number of Americans (or adults in America)

13. $\bar{x} = \dfrac{2+5+3+0+0+(-3)+(-7)+1}{8} = \dfrac{1}{8} = 0.125^\circ$ C; median $= \tilde{x} = \dfrac{0+1}{2} = 0.5^\circ$ C

   (sum of the middle two numbers divided by 2); mode $= 0^\circ$ C

15. $\bar{x} = \dfrac{(4\times 4)+(3\times 3)+(3\times 3)}{10} = \dfrac{34}{10} = 3.4$

   GPA $= 3.4$

17. Various explanations are possible, but the value is probably a median.

19. $\dfrac{(77.6\times 5)+x}{6} = 83.0$

   $(77.6 \times 5) + x = 6(83.0)$

   $388 + x = 498$

   $x = 110$ (Based on a 100 point scale, she cannot attain an average of 83.)

21. a. $\bar{x} = \dfrac{1+2+3+4+4+7+11}{7} = \dfrac{32}{7} = 4.57$

   mean $= 4.57$ median $= 4$ mode $= 4$

   b. $\bar{x} = \dfrac{1+2+3+4+4+10+11}{7} = \dfrac{35}{7} = 5$

   The mean becomes 5 and the others remain unchanged.

23. original salaries: $\bar{x} = \dfrac{\$25{,}000 + \$30{,}000 + \$35{,}000}{3} = \dfrac{\$90{,}000}{3} = \$30{,}000$

   after the raise: $\bar{x} = \dfrac{\$27{,}000 + \$32{,}000 + \$37{,}000}{3} = \dfrac{\$96{,}000}{3} = \$32{,}000$

   The mean would increase by \$2000.

25. No because the one extremely large salary will skew the mean to a higher number than the typical resident makes.

27. $\dfrac{x}{5} = 8$ where $x = $ sum of five numbers

   $x = 40$

   $\dfrac{40-y}{4} = 7$ where $y = $ number removed from list

   $40 - y = 7(4)$

   $40 - y = 28$

   $40 - 28 = y$ The number that was removed is 12.

29. This requires that a class survey be done to determine the mode, mean and median.

Practice Set 8.5

1. All data items are identical.

3. $\bar{x} = \dfrac{0+1+2+3+4+5+6+7+8+9}{10} = \dfrac{45}{10} = 4.5, R = 9-0 = 9, s = 3.0$

5. $\bar{x} = \dfrac{0+0+2+5+6+9+11+15+18+18}{10} = \dfrac{84}{10} = 8.4, \ R = 18-0 = 18, s = 6.9$

7. $\bar{x} = \dfrac{6+6+10+12+3+5}{6} = \dfrac{42}{6} = 7, \ R = 12-3 = 9, \ s = 3.3$

9. $\bar{x} = \dfrac{4+0+3+6+9+12+2+3+4+7}{10} = \dfrac{50}{10} = 5, R = 12-0 = 12, s = 3.6$

11. Section 1: $R = 90 - 10 = 80$; $s = 24$
    Section 2: $R = 100 - 20 = 80$; $s = 27$
    Equal values for R might lead you to conclude that the sets were identical but, the different standard deviation values tell you that the sets are really different.

13. The smaller standard deviation for the Forever batteries (5 hours) indicates that their lifetime is much more likely to be close to the average of 155 hours. The larger standard deviation for the Enduring brand (20 hours) indicates that their lifetimes will vary more from the mean of 155 hours.

15. A small standard deviation is desirable and would indicate a consistent packing process and therefore, more consistency in the amount of chips in each bag.

17. a. Original Salaries: $\bar{x} = \dfrac{20{,}000 + 25{,}000 + 30{,}000}{3} = \dfrac{75{,}000}{3} = \$25{,}000$; $s = 5{,}000$

    With $1000 Raise: $\bar{x} = \dfrac{21{,}000 + 26{,}000 + 31{,}000}{3} = \dfrac{78{,}000}{3} = \$26{,}000$; $s = 5{,}000$

    The mean will rise by $1000 but, since the relative positions of the salaries is not changed, the standard deviation will not change.

    b. With 3% Raise: $\bar{x} = \dfrac{20{,}600 + 25{,}750 + 30{,}900}{3} = \dfrac{77{,}250}{3} = \$25{,}750$; $s = 5{,}150$

    Because 3% of each salary is a different amount, different amounts will be added to each salary so, both the mean and standard deviation will increase by 3%.

174

19. The mean will increase by the amount added but the range and standard deviation will not change. See problem 17a for an example.

21. Because everyone received the same raise, the standard deviation remains the same since the differences among the data as compared to the mean remain unchanged.

Practice Set 8.6

1. Since 830 hours is exactly one standard deviation above the mean ($750 + 80 = 830$ hours), based on the table of z-scores, 34.1% of the bulbs will last between 750 and 830 hours.

3. Since 670 hours is exactly one standard deviation below the mean ($750 - 80 = 670$ hours), based on the table of z-scores, $0.5 - 0.341 = 0.159$ or 15.9% of the bulbs will burn out before being used 670 hours.

5. $z = \dfrac{15 - 15.5}{0.5} = \dfrac{-0.5}{0.5} = -1.00$; Based on the table of z-scores, 34.1% of values are between the mean (15.5 in) and $1s$ below average. To get the % of men that have a neck size of 15 in or larger, we add 34.1% to 50% (the percentage above the mean) to get a total of 84.1% of men with neck sizes greater than 15.5 in.

7. Using the percent calculated in problem 6, we know that 95.4% of men have neck sizes between 14.5 in. and 16.5 in. In a group of 75,000 men, we would expect that $(95.4\%)(75,000) = 0.954 \times 75,000 = 71,550$ of the men would have necks in this size range.

9. $z = \dfrac{350 - 500}{100} = \dfrac{-150}{100} = -1.50$; Based on the table of z-scores, 0.433 or 43.3% of values are between the mean and a z-score of -1.50. Therefore, $0.5 - 0.433 = 0.067$ or 6.7% of values are below the z-score -1.50. Therefore, 6.7% of students are expected to score less than 350 on the SAT Math.

11. The area from the mean to the 0.01 area mark $= 0.5 - 0.01 = 0.49$. Look in the Table of z-Scores and locate the z-score associated with A = 0.490. You will find $z = 2.31$. Because we are looking at the lowest 1%, that is below the mean so the z-score will be negative. Substitute the known values into the z-score formula.

$-2.31 = \dfrac{x - 1200}{120}$

$(-2.31)(120) = x - 1200$

$$-277.2 = x - 1200$$
$$922.8 = x$$

So 1% of the bulbs will burn out before approximately 920 hours of use.

13. Because the Table of z-scores gives areas from the mean to the z-score, if we want to find the area below 95%, we must find the z-score for the area that is between the mean and the upper 45% of the curve. Look in the Table of z-Scores and locate the z-score associated with A = 0.4500. You will find z = 1.64. Because this value is above the mean, the z-score will be positive. Substitute the known values into the z-score formula.

$$1.64 = \frac{x - 1200}{120}$$
$$(1.64)(120) = x - 1200$$
$$196.8 = x - 1200$$
$$1396.8 = x$$

So 95% of the bulbs will burn out before approximately 1397 hours of use.

15. $z = \dfrac{22 - 20}{2.0} = \dfrac{2}{2} = 1.00$; Based on the table of z-scores, 0.341 or 34.1% of values are between the mean and this z-score. Therefore, $0.5 - 0.341 = 0.159$ or 15.9% of values are above this score. Therefore, there is a 15.9% probability that a child will watch TV for more than 22 hours per week.

17. $z = \dfrac{15 - 20}{2.0} = \dfrac{-5}{2} = -2.50$ and $z = \dfrac{25 - 20}{2.0} = \dfrac{5}{2} = 2.50$. Based on the Table of z-Scores, the area between the mean and z = 2.50 is 0.494. Since the curve is symmetrical, the area between the mean and z = -2.50 is also 0.494, for a total proportion of 0.988 or about 99%. Therefore, there is a 99% probability that a child watches TV for between 15 and 25 hours per week.

19.
$$\bar{x} = \frac{37 + 52 + 25 + 48 + 26 + 41 + 22 + 15 + 52 + 40 + 50 + 16 + 58 + 59 + 51 + 26 + 39 + 12 + 37 + 20}{20} =$$

$$\frac{726}{20} = 36.3; \; s = 15.10$$

21. $z = \dfrac{65 - 72}{5} = \dfrac{-7}{5} = -1.40$; Based on the table of z-scores, 0.419 or 41.9% of values are between the mean and a z-score of -1.40. Therefore, $0.5 + 0.419 = 0.919$ or 91.9% of values are above the z-score -1.40. Therefore, 91.9% of males are taller than 65 inches.

23. $z = \dfrac{64 - 72}{5} = \dfrac{-8}{5} = -1.60$ and $z = \dfrac{74 - 72}{5} = \dfrac{2}{5} = 0.40$. Based on the Table of z -Scores, the area between the mean and z = -1.60 is 0.445. The area between the mean and 0.40 is 0.155. Therefore, the total proportion between these two z-scores is 0.445 + 0.155 = 0.600 and 60% of males will have heights between 64 and 74 in.

25. Based on the definition of percentile, 85% will have scored lower than you.

27. Ray's $z = \dfrac{82 - 65}{12} = \dfrac{17}{12} = 1.42$; Susan's $z = \dfrac{78 - 65}{10} = \dfrac{13}{10} = 1.30$; Ray would receive a higher grade because his grade is farther above the mean average for his class than Susan's is for hers.

29. $z = \dfrac{82 - 78}{5} = \dfrac{4}{5} = 0.80$

31. A negative z-score indicates that the grade would be lower than the mean of 78.

$$-1.6 = \dfrac{x - 78}{5}$$
$$(-1.6)(5) = x - 78$$
$$-8 = x - 78$$
$$70 = x \qquad \text{The grade would be 70.}$$

33. $z = \dfrac{110 - 100}{15} = \dfrac{10}{15} = 0.67$; The proportion of area between the mean and the z-score 0.67 is 0.249. Therefore, the percent of IQs below 110 will include the 50% below the mean of 100 and the 24.9% between 100 and 110 for a total of 74.9% or 75%.

35. The fifth percentile means that only 5% scored lower than this score. The z-score for 5% (0.050) below the mean = -1.64.

$$-1.64 = \dfrac{x - 100}{15}$$
$$(-1.64)(15) = x - 100$$
$$-24.6 = x - 100$$
$$75.4 = x \qquad \text{The IQ score of 75 corresponds to the 5}^{\text{th}} \text{ percentile.}$$

37. 95% in the middle means 47.5% above the mean and below the mean. Therefore, we look in the table for the z-score associated with an area $(A) = 0.475$. This value is 1.96. Since our two values will be both above and below the mean, we use +1.96 and -1.96 to find our answers.

$$1.96 = \frac{x-100}{15}$$
$$(1.96)(15) = x - 100$$
$$29.4 = x - 100$$
$$129.4 = x$$

$$-1.96 = \frac{x-100}{15}$$
$$(-1.96)(15) = x - 100$$
$$-29.4 = x - 100$$
$$70.6 = x$$

The scores that separate the middle 95% from the remainder of the scores are 71 and 129.

39. $z = \dfrac{30-40}{11.4} = \dfrac{-10}{11.4} = -0.88$ and $z = \dfrac{50-40}{11.4} = \dfrac{10}{11.4} = 0.88$. Based on the Table of z-Scores, the area between the mean and z = -0.88 is 0.311. The area between the mean and 0.88 is also 0.311. Therefore, the total proportion between these two z-scores is 2(0.311) = 0.622. Therefore, 62.2% of customers use an ATM between 30 and 50 times. Among 5000 customers, we can predict that 62.2% of 5000 or 0.622 × 5000 = 3110 customers will use the ATM between 30 and 50 times.

41. $z = \dfrac{5\,ft.\,10\,in. - 5\,ft.\,5.5\,in.}{2.0} = \dfrac{4.5\,in.}{2.0} = 2.25$; The proportion of area between the mean and the z-score 2.25 is 0.488. Therefore, the percent of women's heights above 5 ft. 10 in. will be calculated by taking the 50% above the mean of 5 ft. 5.5 in. and subtracting the 48.8% from it for a difference of 1.2% .

Practice Set 8.7

1. The older the car, the less it is worth. This is a negative correlation.

3. As a car's weight increases, its gas mileage will decrease. This is a negative correlation.

5. a. Education is the independent variable and income is the dependent variable.
   b. There is a strong positive correlation based on the r-value.
   c. As one's years of education increase, one's income increases.

7. $y = 0.005(600) + 0.40 = 3.4$ GPA

9. $y = 0.002(1060) + 0.667 = 2.787$ GPA

11. a.

Hospital Stay

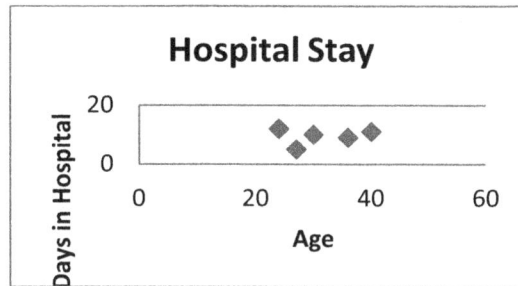

b. There is no obvious linearity in the graph. The r-value (found using the calculator) is 0.158 which indicates weak correlation.

c. $y = 0.065x + 7.35$ (found using the calculator)

d. $y = 0.065(32) + 7.35 = 9.43$ or 9 days; This prediction is not reliable because of the weak correlation between these variables.

13. a.

Baby Weight vs Adult Weight

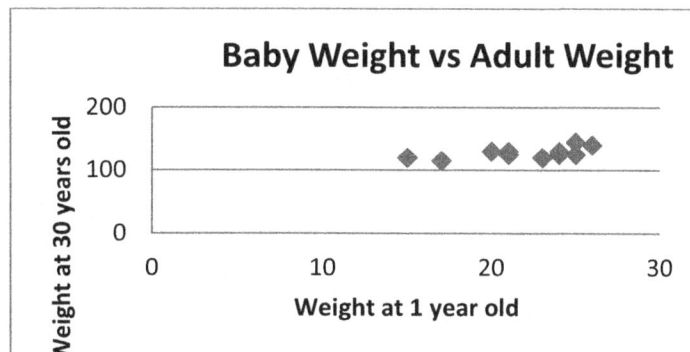

b. There is a moderate correlation (r-value $= 0.64$).

c. $y = 1.6x + 92.5$ (found using the calculator)

d. The correlation coefficient is $r = 0.64$.

e. $y = 1.6(20) + 92.5 = 124.5$ lb

15. a.

Quiz Grades

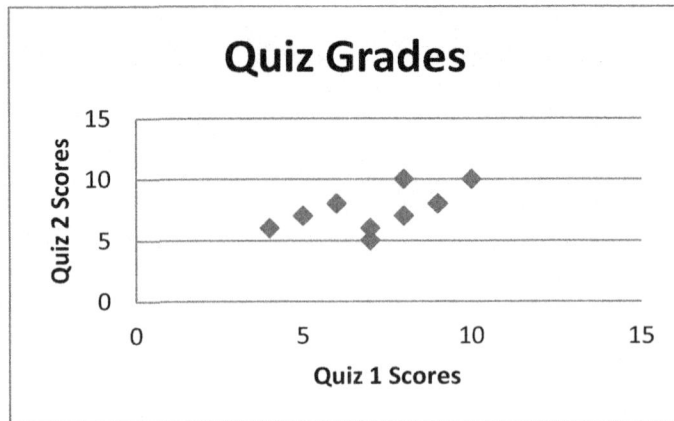

b. There seems to be a moderate correlation with a correlation coefficient $r = 0.55$. It appears in the graph to be a positive linear correlation.

17. a. Yes, there appears to be a positive linear correlation.

Real Estate Sales

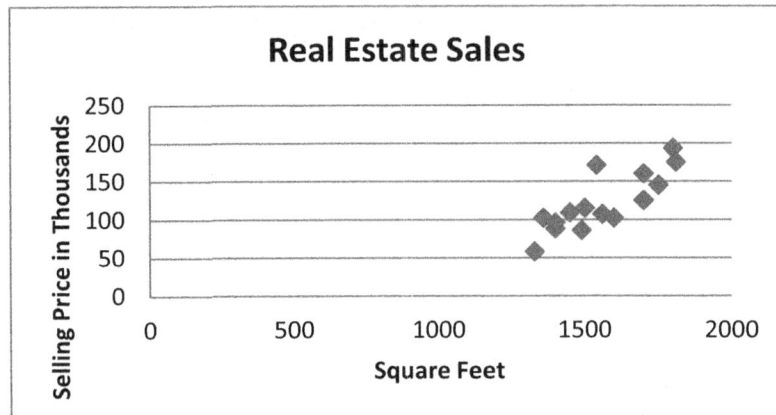

b. The correlation coefficient is $r = 0.82$ indicating a strong correlation.
c. $y = 0.197x - 185$ (Found using the calculator)
d. $y = 0.197(1575) - 185 = 125.275$ or $125,275; This predicted price should be fairly reliable since the correlation between the two variables is strong.

Chapter 8 Review Problems

1. Descriptive statistics makes no predictions or guesses. It summarizes a population of data. Inferential statistics takes descriptive data and uses them to help make "educated guesses" about the data or about the group from which the data were gathered.

2. Quantitative data are data that can be "quantified," i.e., they have numbers and amounts associated with them. Qualitative data are non-numerical in nature.

3. cluster sample; probably not biased if flights chosen randomly from many airports

4. convenience sampling; could be biased if the interviewer chooses only people who seem friendly or who come in during a certain time of day

5. stratified sample; probably not biased if representative samples from each stratum are polled

6. Sample results cannot be applied to the general population since it was not representative.

7. The number seems too precise for a weight loss figure. Also, sample size is not given.

8. sum = 126, n = 9, mean = 14, median = 14, no mode, $R = 8$, $s = 2.7$ (all as determined using a graphing calculator and methods shown in the text)

9. sum = 173, n = 10, mean = 17.3, median = 8.5, mode = 8, $R = 55$, $s = 17.2$ (all as determined using a graphing calculator and methods shown in the text)

10. The data set in problem 9 has more variation since its standard deviation is larger.

11. For the data in problem 8, the mean is the best average because there are no big gaps in the list of numbers. For the data in problem 9, the median is the best average because of the relatively large jump in the data list to get to 57.

12.

Classes	Frequency
13-14	4
11-12	5
9-10	5
7-8	13
5-6	12
3-4	11

13. The histogram is positively skewed.

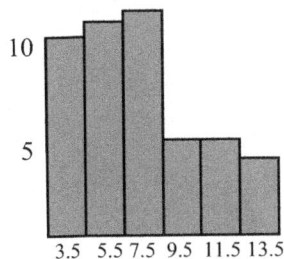

14. $z = \dfrac{37-35}{2} = \dfrac{2}{2} = 1.00$ The area between the mean and this z-score is 0.341. To find the area above this score, subtract 0.341 from 0.5 to give an area of 0.159 which indicates that about 16% of men have a waist size greater than 37 inches.

181

15. The mean average of the survey is 35 in. The mean average is the center of the normal distribution with 50% above and 50% below its value.

16. $z = \dfrac{37-35}{2} = \dfrac{2}{2} = 1.00$; $z = \dfrac{31-35}{2} = \dfrac{-4}{2} = -2.00$; The area between the mean and the z-score of 1.00 is 0.341. The area between the mean and the z-score of -2.00 is 0.477. To find the area between these two scores, add 0.341 and 0.477 to give an area of 0.818 which indicates that about 82% of men have a waist size between 31 inches and 37 inches.

17. math test: $z = \dfrac{88-80}{5} = \dfrac{8}{5} = 1.60$; history test: $z = \dfrac{88-85}{3} = \dfrac{3}{3} = 1.00$. The z-scores indicate that you scored farther above the class average on the math test than the history test. This means that, relatively speaking, you did better on the math test.

18. $z = \dfrac{30,000-45,000}{8200} = \dfrac{-15,000}{8200} = -1.83$; The area between the mean and this z-score is 0.466. To find the area below this score, subtract 0.466 from 0.5 to give an area of 0.034 which indicates the proportion that they expect to replace under this guarantee.

19. $z = \dfrac{40,000-45,000}{8200} = \dfrac{-5,000}{8200} = -0.61$; The area between the mean and this z-score is 0.229. To find the area below this score, subtract 0.229 from 0.5 to give an area of 0.271 which indicates the proportion that they expect to replace under this guarantee.

20. There does not appear to be a strong correlation.

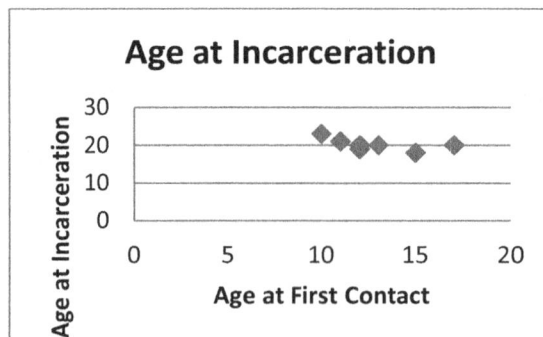

21. $y = -0.40x + 25.3$ (Found using the calculator)
22. $y = -0.40(14) + 25.3 = 19.7$ or about age 20
23. The correlation coefficient is $r = -0.61$ (Found using the calculator)
24. The r-value indicates a moderate negative correlation of these two variables.

182

Chapter 8 Test

1. quantitative because a *count* is a numerical amount
2. qualitative because an *occupation* is non-numerical
3. qualitative because the *name of a television* show is non-numerical
4. Because the number is a percent, you might investigate how many students took the test. It could be a very low number.
5. Data set B because of the unusually large value of 19 as compared to the other data
6. The mean is the best average for data set A, and the median is the best average for data set B because of the extreme value of 19 and how it affects the mean.
7. $n = 11$ and the data sum $= 402$, mean $= \dfrac{402}{11} = 36.5$; median $= 36$ (the data are already sorted so count 6 items from either end); mode $= 36$; $R = 44 - 32 = 12$; $s = 3.9$ by the calculator method shown in the text
8. median $= M$

9.
0	3
1	2 7 0 8 8 9 2 8
2	7 7 2 9 1 6 1 8 9 1 6
3	0 5 8 6 5 8 2 3 2 1 2 3 1 2
4	2 7 1 5 3
5	1

10. It appears to be approximately normal.

11. $z = \dfrac{37 - 41}{4} = \dfrac{-4}{4} = -1.00$; The area between the mean and this z-score is 0.341. To find the area below this score, subtract 0.341 from 0.5 to give an area of 0.159 which indicates that about 16% of incoming students must take this course.

12. $z = \dfrac{65 - 80}{5} = \dfrac{-15}{5} = -3.00$

13. $1.75 = \dfrac{x - 80}{5}$

$(1.75)(5) = x - 80$

$8.75 = x - 80$

$88.75 = x$

14. $z = \dfrac{70 - 80}{5} = \dfrac{-10}{5} = -2.00$; The area between the mean and this z-score is 0.477. To calculate the entire area higher than the value of 70, add $0.477 + 0.500$ to get 0.977 which is the proportion of students who passed.

15. $z = \dfrac{89 - 80}{5} = \dfrac{9}{5} = 1.80$; The area between the mean and this z-score is 0.464. This represents the proportion of students who scored between 80 (the mean) and 89.

16. A z-score is the number of standard deviations a value is above or below the mean average . The z-score for your height would be 2.00.

17. $z = \dfrac{60 - 64}{3} = \dfrac{-4}{3} = -1.33$; The area between the mean and this z-score is 0.408. To find the area below this score, subtract 0.408 from 0.5 to give an area of 0.092 which indicates the proportion that they expect to replace under this guarantee.

18. $y = 4(69) - 130 = 146$ lb

19. $r = 0.80$ (Found using calculator); This value indicates a fairly strong correlation between an employee's age and his salary.

20.

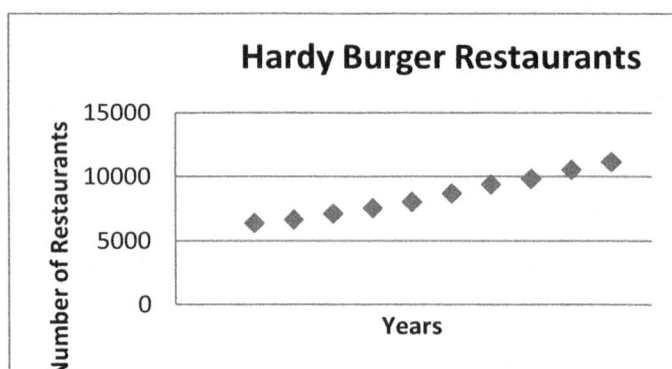

21. $y = 544.4x - 1081627.2$ (Found using the calculator)

22. $y = 544.4(2017) - 1,081,627.2 = 16,427.6$ or 16,428 predicted to be open in 2017.

Graphing Calculator Quick References {TI-84 ®}

What follows is a calculator supplement for *Fundamentals of Algebraic Modeling*. It is intended to give both the instructor and the student some basic instructions for use of a graphing calculator. These instructions are specifically for the Texas Instruments TI-83+® but are easily adaptable to the TI-84® or other similar calculators. Only operations necessary for use in the text are included.

Graphing Points With The 'Stat-Editor'

Suppose that you wished to graph the following set of individual points:

$$\{(2,4), (0,2), (-5,7), (8,0)\}.$$

There are several ways to accomplish this on the TI-83 PLUS but, since the LIST feature is used in several other applications, we will use it here.

1. Since we are going to graph these points, first we need to be sure that the calculator's STAT PLOT is cleared out so that no "old" equations will be plotted along with our points. To do this press Y= . If any equations are listed in Y_1 or any of the other equation slots, use the *arrow keys* to move the cursor so that the equation is highlighted, then press CLEAR. Repeat this until no equations are listed.

2. Press **STAT** and then **1** for the editor

3. If there are already some numbers in the lists, first clear them by doing the following:
    A. press the *up arrow*, σ, until the cursor is on L_1
    B. press **CLEAR** and then **ENTER** to erase list one
    C. repeat this for any other lists that contain numbers

4. Using the *arrow keys*, move the cursor into the location for the first number in L_1

5. L_1 will hold all of the *x* values from your number pairs and L_2 will contain the *y* values. Type

Calculator.lnk   in the first *x* and then press **ENTER**. The cursor will move down so that you may now enter the next *x* value by the same method. After entering the last *x* value, use the *arrow keys* to move over to L_2 and enter the *y* values.

6. After entering all values, your screen should look like this:

L₁	L₂	L₃
2	4	------
0	2	
-5	7	
8	0	
------	------	

7. Now press **2ⁿᵈ** and then **Y=** to get to **STAT PLOT**.

8. Press **ENTER** on number 1.

9. Move the cursor with the *arrow keys* to the word **ON** and then press **ENTER** so that the shaded box is on **ON**.
10. Using the *arrow keys* move down to the first **TYPE** of graph (dotted) and then press **ENTER**.

11. Make sure that the **Xlist** has L_1 highlighted and the **Ylist** has L_2 highlighted. If they are not, then use the *arrow keys* to move the cursor to the proper list and then highlight it by pressing the **ENTER** key.

12. Press **ZOOM** and **9** to see the graph of the points. **ZOOM 9** will automatically adjust the **WINDOW** settings to fit the points you entered into the lists.

Solving Equations By Graphing

Let's start with a simple linear equation as an example. Suppose you were asked to solve the following:

$$2x + 1 = 9$$

1. Press **Y=** as if you were going to graph a regular function.

2. After clearing out any equations that may have been entered previously, type in the first side of the equation as Y_1 and the other side as Y_2. Your screen should now say:

$$Y_1 = 2x + 1$$
$$Y_2 = 9$$

3. Press **GRAPH**. You should see the two lines crossing each other. If not, press **ZOOM** and then **6** (for standard zoom). For other equations you may need to adjust the **WINDOW** settings.

4. Press **2nd** and **CALC** (over the **TRACE** button.)

5. Press **5** for intersection. Move the cursor, using the *right and left arrow keys*, so that it is as close as possible to the point at which the two lines intersect.

6. Press **ENTER, ENTER, ENTER** . (i.e., press the ENTER key three times in succession.)

7. The x value shown on the screen is the solution to the original equation. In this case you should get $x = 4$.

8. Now try to solve $2x + 4 = x + 1$. $\{x = -3\}$

Graphing Piece-Wise Functions

Suppose that you have a function such as the following to graph:

$$f(x) = \{x \text{ if } x < 0 \text{ and } x^2 \text{ if } x \geq 0\}.$$

This type of function is called "piece-wise" since there is no direct connection between values of the function for values of x that are less than zero and those that are equal to zero and greater.

1. Press **Y=** and first clear out any equations that may have been entered previously.

2. For equation Y_1 type in **x/(x<0)** as follows:
   a. press **XTΘ**
   b. press the ÷ key
   c. press **(**
   d. press **XTΘ**
   e. to get to the inequality symbols press **2nd** and **MATH** to get to the **TEST** menu, then press **5** and the '<' symbol should appear as part of Y_1
   f. press **)**

3. Press the *down arrow* to move down to the second equation.

4. Now enter the equation for Y_2 in the same manner. You should type in $x^2/(x \geq 0)$ as follows:
   XTΘ, x^2, ÷, (, XTΘ, 2nd, MATH, 4, 0,)

5. Press **GRAPH** to see this piece-wise function's graph.

6. Graph the following function:

$$f(x) = \{x \text{ if } x < 2 \text{ and } 2x \text{ if } x \geq 2\}.$$

Made in the USA
Columbia, SC
08 August 2023